RECENT PROGRESS IN DIFFERENTIAL GEOMETRY AND ITS RELATED FIELDS

RECENT PROGRESS IN DIFFERENTIAL GEOMETRY AND ITS RELATED FIELDS

Proceedings of the 2nd International Colloquium on Differential Geometry and its Related Fields

Veliko Tarnovo, Bulgaria   6 – 10 September 2010

editors

Toshiaki Adachi
Nagoya Institute of Technology, Japan

Hideya Hashimoto
Meijo University, Japan

Milen J Hristov
St Cyril and St Methodius University of Veliko Tarnovo, Bulgaria

World Scientific

NEW JERSEY · LONDON · SINGAPORE · BEIJING · SHANGHAI · HONG KONG · TAIPEI · CHENNAI

Published by

World Scientific Publishing Co. Pte. Ltd.

5 Toh Tuck Link, Singapore 596224

USA office: 27 Warren Street, Suite 401-402, Hackensack, NJ 07601

UK office: 57 Shelton Street, Covent Garden, London WC2H 9HE

British Library Cataloguing-in-Publication Data
A catalogue record for this book is available from the British Library.

ISBN-13 978-981-4355-46-9
ISBN-10 981-4355-46-1

Desk Editor: Lai Fun Kwong
Cover Designer: Hui Chee Lim

Photo Location: Tsarevets, Veliko Tarnovo

Printed in Singapore.

PREFACE

The *2nd International Colloquium on Differential Geometry and its Related Fields* (ICDG2010) was held at St. Cyril and St. Methodius University of Veliko Tarnovo, Bulgaria, during the period of 6–10 September, 2010.

This International Colloquium was mooted to provide opportunities for mathematicians both in the western and eastern countries to interchange ideas and discuss mathematics quite freely and in a homely atmosphere not only in the area of differential geometry but also in other fields of sciences. This volume contains selected research papers which treat geometric structures, concrete Lie group theory and information geometry. The editors expect these articles to provide significant information for researchers, serve as a good guide for graduate students and confer a benefit upon further development of studies in differential geometry, complex analysis and some related fields.

This academic program is also based on the agreement of academic exchange and cooperation between St. Cyril and St. Methodius University of Veliko Tarnovo and Department of Computer Science of Engineering, Nagoya Institute of Technology. We gratefully acknowledge the staff of Veliko Tarnovo University for their help in coordinating this colloquium and their warm hospitality. We would like to express our sincere gratitude to all the referees for their assistance. We would also like to acknowledge the partial financial support from Daiko Foundations and Meijo University.

The Editors
May 5, 2011

The 2nd International Colloquium on Differential Geometry and its Related Fields

6–10 September, 2010 – Veliko Tarnovo, Bulgaria

ORGANIZING COMMITTEE

T. Adachi	– Nagoya Institute of Technology, Nagoya, Japan
S. H. Bouyuklieva	– St. Cyril and St. Methodius University of Veliko Tarnovo, Veliko Tarnovo, Bulgaria
H. Hashimoto	– Meijo University, Nagoya, Japan
M. J. Hristov	– St. Cyril and St. Methodius University of Veliko Tarnovo, Veliko Tarnovo, Bulgaria

SCIENTIFIC ADVISORY COMMITTEE

G. Ganchev	– Bulgarian Academy of Sciences, Sofia, Bulgaria
K. Sekigawa	– Niigata University, Niigata, Japan
T. Sunada	– Meiji University, Tokyo, Japan

PRESENTATIONS

1. **Stefka Bouyuklieva** (Univ. Veliko Tarnovo),
   Singly-even self-dual codes and their shadows

2. **Iliya Bouyukliev** (Bulgarian Acad. Sci.),
   Codes, sets of points in finite projective geometries and dual transform

3. **Marija Djumalieva-Stoeva** (Univ. Veliko Tarnovo),
   Representation of the equivalence problem for sets of points in projective geometries

4. **Stela Zhelezova** (Bulgarian Acad. Sci.),
   t-parallelisms of PG(5,2) with automorphisms of order 31

5. **Zlatko Varbanov** (Univ. Veliko Tarnovo),
   Techniques for bit manipulation and their applications

6. **Emilian Petkov** (Univ. Veliko Tarnovo),
   On NURBS curves and surfaces in computer graphics

7. **Hiroshi Matsuzoe** (Nagoya Inst. Tech.),
   Geometry of nonextensive generalized entropies

8. **Hideya Hashimoto** (Meijo Univ.),
   On automorphism groups of 6-submanifolds of the octonoins

9. **Misa Ohashi** (Meijo Univ.),
   On generalized cylindrical helices and Lagrangian surfaces of $\mathbb{R}^4$

10. **Yusuke Sakane** (Osaka Univ.),
    Invariant Einstein metrics on generalized flag manifolds

11. **Georgi Ganchev** (Bulgarian Acad. Sci.),
    A classification of Riemannian manifolds of quasi-constant sectional curvatures and conformally flat hypersurfaces in Euclidean or minkowski space

12. **Milen Hristov** (Univ. Veliko Tarnovo),
    Bézier-type almost complex structures on quaternionic Kähler manifolds

13. **Tuya Bao** (Nagoya Inst. Tech. & Inner Mongolia Univ. Nationalities),
    Length spectrum of circular trajectories on geodesic spheres in a nonflat complex space form

14. **Galia Nakova** (Univ. Veliko Tarnovo),
    Lightlike submanifolds of almost complex manifolds with Norden metric

15. **Nikolaj Ivanov** (Univ. Veliko Tarnovo),
    String theory and K-theory

16. **Mancho Manev** (Plovdiv Univ.),
    On connections with totally skew-symmetric torsion on manifolds with additional tensor structures and indefinite metrices

17. **Andrea Loi** (Univ. Cagliari),
    Kaehler immersions of homogeneous Kaehler manifolds into complex space forms

18. **Michela Zedda** (Univ. Cagliari),
    Balanced metrics on Hartogs domains

19. **Roberto Mossa** (Univ. Cagliari),
    The diastatic exponential

20. **Velichka Milousheva** (Bulgarian Acad. Sci.),
    An invariant theory of surfaces in the four-dimensional Euclidean or Minkowski space

21. **Daniele Zuddas** (Univ. Cagliari),
    Lefschetz fibration structures on 4-dimensional 2-handlebodies

22. **Toshiaki Adachi** (Nagoya Inst. Tech.),
    A theorem of Hadamard-Cartan type for Kähler magnetic fields

23. **Emil Arsov** (Univ. Veliko Tarnovo),
    Animation by some conformal transformations

24. **Plamen Assenov** (Univ. Veliko Tarnovo),
    Animation by affine transformations of Bézier curves

CONTENTS

Proceedings of the 2nd International
Colloquium on Differential Geometry
and its Related Fields
Veliko Tarnovo, September 6–10, 2010

HOMOGENEOUS EINSTEIN METRICS
ON GENERALIZED FLAG MANIFOLDS
$Sp(n)/(U(p) \times U(q) \times Sp(n - p - q))$

Andreas ARVANITOYEORGOS* and Ioannis CHRYSIKOS**

Department of Mathematics, University of Patras,
GR-26500 Rion, Greece
*E-mail: arvanito@math.upatras.gr
**E-mail: xrysikos@master.math.upatras.gr

Yusuke SAKANE

Department of Pure and Applied Mathematics,
Graduate School of Information Science and Technology, Osaka University,
Osaka 560-043, Japan
E-mail: sakane@math.sci.osaka-u.ac.jp

We construct the Einstein equation for an invariant Riemannian metric on
generalized flag manifolds $Sp(n)/(U(p) \times U(q) \times Sp(n-p-q))$. By computing
a Gröbner basis for a system of polynomials on six variables, we prove that
the generalized flag manifolds $Sp(3)/(U(1) \times U(1) \times Sp(1))$, $Sp(4)/(U(1) \times U(1) \times Sp(2))$ and $Sp(4)/(U(2) \times U(1) \times Sp(1))$ admit exactly three, six and
two non-Kähler invariant Einstein metrics up to isometry, respectively.

Keywords: Homogeneous Einstein metric, Generalized flag manifold, Gröbner
basis.

1. Introduction

A Riemannian manifold (M, g) is called Einstein if the Ricci tensor r of the
metric g satisfies $r = \lambda g$ for some $\lambda \in \mathbb{R}$. In this paper we discuss homogeneous Einstein metrics on generalized flag manifolds. A generalized flag
manifold is an adjoint orbit of a compact semisimple Lie group G, or equiv-
alently a compact homogeneous space of the form $M = G/K = G/C(S)$,
where $C(S)$ is the centralizer of a torus S in G. Einstein metrics on generalized flag manifolds have been studied by several authors (Alekseevsky,
Arvanitoyeorgos, Chrysikos, Kimura, Negreiros, Sakane).

Generalized flag manifolds G/K admit a finite number of G-invariant
complex structures and invariant Kähler Einstein metrics on general-

ized flag manifolds were investigated by Alekseevsky and Perelomov[2]. The problem of finding non-Kähler Einstein metrics on generalized flag manifolds was initially studied by Alekseevsky[1] and Arvanitoyeorgos[4]. Kimura[13] studied all invariant Einstein metrics for generalized flag manifolds with three isotropy summands. (See also [3], [6].)

In recent works [5], [7] and [8], all invariant Einstein metrics were found for generalized flag manifolds with four isotropy summands. Moreover, all invariant Einstein metrics on the generalized flag manifold $SO(7)/(U(1) \times U(1) \times SO(3))$ with five isotropy summands and the exceptional full flag manifold G_2/T with six isotropy summands are found in [11] and [9] respectively.

In this paper we deal with generalized flag manifolds $Sp(n)/(U(p) \times U(q) \times Sp(n-p-q))$ with six isotropy summands. We give explicit expressions of Ricci tensor of the invariant metrics on these generalized flag manifolds. To give such expressions we use a Riemannian submersion from a generalized flag manifold to the other generalized flag manifold with totally geodesic fibers and Kähler Einstein metrics on the generalized flag manifolds.

Our main results are the following:

Theorem 1.1. *The generalized flag manifold $Sp(3)/(U(1) \times U(1) \times Sp(1))$ admits exactly one Kähler Einstein metric and three non-Kähler Einstein metrics up to isomeries. These Einstein metrics are given by (25), (23) and (26).*

Theorem 1.2. *The generalized flag manifold $Sp(4)/(U(1) \times U(1) \times Sp(2))$ admits exactly one Kähler Einstein metric and six non-Kähler Einstein metrics up to isomeries. These Einstein metrics are given by (33), (32), (34), (36) and (37).*

Theorem 1.3. *The generalized flag manifold $Sp(4)/(U(2) \times U(1) \times Sp(1))$ admits exactly two Kähler Einstein metrics and two non-Kähler Einstein metrics up to isomeries. These Einstein metrics are given by (39) and (40).*

2. Ricci tensor of a compact homogeneous space G/K

Let G be a compact semi-simple Lie group, K a connected closed subgroup of G and let $\mathfrak{g}$ be the Lie algebra of G, $\mathfrak{k}$ the Lie algebra of K. Note that the Killing form of a compact Lie algebra $\mathfrak{g}$ is negative definite. We set $B = -$ Killing form. Then B is an $\mathrm{Ad}(G)$-invariant inner product on $\mathfrak{g}$.

Let $\mathfrak{m}$ be the orthogonal complement of $\mathfrak{k}$ in $\mathfrak{g}$ with respect to B. Then we have $\mathfrak{g} = \mathfrak{k} \oplus \mathfrak{m}$, $[\mathfrak{k}, \mathfrak{m}] \subset \mathfrak{m}$ and a decomposition of $\mathfrak{m}$ into irreducible $\mathrm{Ad}(K)$-modules:

$$\mathfrak{m} = \mathfrak{m}_1 \oplus \cdots \oplus \mathfrak{m}_q.$$

We assume that $\mathrm{Ad}(K)$-modules $\mathfrak{m}_j$ $(j = 1, \cdots, q)$ are mutually non-equivalent. Then a G-invariant metric on G/K can be written as

$$\langle \, , \, \rangle = x_1 B|_{\mathfrak{m}_1} + \cdots + x_q B|_{\mathfrak{m}_q}, \tag{1}$$

for positive real numbers $x_1, \cdots, x_q$.

Note that G-invariant symmetric covariant 2-tensors on G/K are the same form as the metrics. In particular, the Ricci tensor r of a G-invariant Riemannian metric on G/K is of the same form as (1).

Let $\{e_\alpha\}$ be a B-orthonormal basis adapted to the decomposition of $\mathfrak{m}$, i.e., $e_\alpha \in \mathfrak{m}_i$ for some i, and $\alpha < \beta$ if $i < j$ (with $e_\alpha \in \mathfrak{m}_i$ and $e_\beta \in \mathfrak{m}_j$). We put $A_{\alpha\beta}^\gamma = B([e_\alpha, e_\beta], e_\gamma)$, so that $[e_\alpha, e_\beta] = \sum_\gamma A_{\alpha\beta}^\gamma e_\gamma$, and set $\begin{bmatrix} k \\ ij \end{bmatrix} = \sum (A_{\alpha\beta}^\gamma)^2$, where the sum is taken over all indices α, β, γ with $e_\alpha \in \mathfrak{m}_i$, $e_\beta \in \mathfrak{m}_j$, $e_\gamma \in \mathfrak{m}_k$ (cf. [18]). Then, the positive number $\begin{bmatrix} k \\ ij \end{bmatrix}$ is independent of the B-orthonormal bases chosen for $\mathfrak{m}_i, \mathfrak{m}_j, \mathfrak{m}_k$, and

$$\begin{bmatrix} k \\ ij \end{bmatrix} = \begin{bmatrix} k \\ ji \end{bmatrix} = \begin{bmatrix} j \\ ki \end{bmatrix}. \tag{2}$$

Let $d_k = \dim \mathfrak{m}_k$. Then we have

Lemma 2.1. [16] *The components $r_1, \cdots, r_q$ of Ricci tensor r of the metric $\langle \, , \, \rangle$ of the form* (1) *on G/K are given by*

$$r_k = \frac{1}{2x_k} + \frac{1}{4d_k} \sum_{j,i} \frac{x_k}{x_j x_i} \begin{bmatrix} k \\ ji \end{bmatrix} - \frac{1}{2d_k} \sum_{j,i} \frac{x_j}{x_k x_i} \begin{bmatrix} j \\ ki \end{bmatrix} \quad (k = 1, \cdots, q) \tag{3}$$

where the sum is taken over $i, j = 1, \cdots, q$.

3. Riemannian submersion

Let G be a compact semi-simple Lie group and K, L two closed subgroups of G with $K \subset L$. Then we have a natural fibration $\pi : G/K \to G/L$ with fiber L/K.

Let $\mathfrak{p}$ be the orthogonal complement of $\mathfrak{l}$ in $\mathfrak{g}$ with respect to B, and $\mathfrak{n}$ be the orthogonal complement of $\mathfrak{k}$ in $\mathfrak{l}$. Then we have $\mathfrak{g} = \mathfrak{l} \oplus \mathfrak{p} = \mathfrak{k} \oplus \mathfrak{n} \oplus \mathfrak{p}$.

An $\mathrm{Ad}_G(L)$-invariant scalar product on $\mathfrak{p}$ defines a G-invariant metric $\check{g}$ on G/L, and an $\mathrm{Ad}_L(K)$-invariant scalar product on $\mathfrak{n}$ defines an L-invariant metric $\hat{g}$ on L/K. The orthogonal direct sum for these scalar products on $\mathfrak{n} \oplus \mathfrak{p}$ defines a G-invariant metric g on G/K.

Theorem 3.1. ([10] p. 257) *The map π is a Riemannian submersion from $(G/K, g)$ to $(G/L, \check{g})$ with totally geodesic fibers isometric to $(L/K, \hat{g})$.*

Note that $\mathfrak{n}$ is the vertical subspace of the submersion and $\mathfrak{p}$ is the horizontal subspace.

For a Riemannian submersion, O'Neill [15] has introduced two tensors A and T. Since the fibers are totally geodesic in our case, we have $T = 0$. We also have

$$A_X Y = \frac{1}{2}[X, \ Y]_\mathfrak{n} \quad \text{for } X, Y \in \mathfrak{p}.$$

Let $\{X_i\}$ be an orthonormal basis of $\mathfrak{p}$ and $\{U_j\}$ an orthonormal basis of $\mathfrak{n}$. We put for $X, Y \in \mathfrak{p}$, $g(A_X, \ A_Y) = \sum_i g(A_X X_i, A_Y X_i)$. Then we have

$$g(A_X, \ A_Y) = \frac{1}{4}\sum_i \hat{g}([X, \ X_i]_\mathfrak{n}, \ [Y, \ X_i]_\mathfrak{n}).$$

Let $r, \check{r}$ be the Ricci tensor of the metric $g, \check{g}$ respectively. Then we have

$$r(X,Y) = \check{r}(X,Y) - 2g(A_X, A_Y) \quad \text{for } X, Y \in \mathfrak{p}.$$

Let

$$\mathfrak{p} = \mathfrak{p}_1 \oplus \cdots \oplus \mathfrak{p}_\ell, \quad \mathfrak{n} = \mathfrak{n}_1 \oplus \cdots \oplus \mathfrak{n}_s$$

be a decomposition of $\mathfrak{p}$ into irreducible $\mathrm{Ad}(L)$-modules and a decomposition of $\mathfrak{n}$ into irreducible $\mathrm{Ad}(K)$-modules respectively. We assume that $\mathrm{Ad}(L)$-modules $\mathfrak{p}_j$ $(j = 1, \cdots, \ell)$ are mutually non-equivalent. Note that each irreducible component $\mathfrak{p}_j$ as $\mathrm{Ad}(L)$-module can be decomposed into irreducible $\mathrm{Ad}(K)$-modules. To compute the values $\begin{bmatrix} k \\ ij \end{bmatrix}$ for G/K, we use information from the Riemannian submersion $\pi : (G/K, g) \to (G/L, \check{g})$ with totally geodesic fibers isometric to $(L/K, \hat{g})$. We consider a G-invariant metric on G/K defined by a Riemannian submersion $\pi : (G/K, g) \to (G/L, \check{g})$ of the form

$$g = y_1 B|_{\mathfrak{p}_1} + \cdots + y_\ell B|_{\mathfrak{p}_\ell} + z_1 B|_{\mathfrak{n}_1} + \cdots + z_s B|_{\mathfrak{n}_s} \tag{4}$$

for positive real numbers $y_1, \cdots, y_\ell, z_1, \cdots, z_s$.

We decompose each irreducible component $\mathfrak{p}_j$ into irreducible $\mathrm{Ad}(K)$-modules:

$$\mathfrak{p}_j = \mathfrak{m}_{j,1} \oplus \cdots \oplus \mathfrak{m}_{j,k_j}.$$

As before we assume that $\mathrm{Ad}(K)$-modules $\mathfrak{m}_{j,t}$ ($j = 1, \cdots, \ell$, $t = 1, \cdots, k_j$) are mutually non-equivalent. Note that the metric of the form (4) can be written as

$$g = y_1 \sum_{t=1}^{k_1} B|_{\mathfrak{m}_{1,t}} + \cdots + y_\ell \sum_{t=1}^{k_\ell} B|_{\mathfrak{m}_{\ell,t}} + z_1 B|_{\mathfrak{n}_1} + \cdots + z_s B|_{\mathfrak{n}_s} \qquad (5)$$

and this is a special case of the metric of the form (1).

Lemma 3.1. *Let* $d_{j,t} = \dim \mathfrak{m}_{j,t}$. *The components* $r_{(j,t)}$ ($j = 1, \cdots, \ell$, $t = 1, \cdots, k_j$) *of Ricci tensor* r *for the metric* (5) *on* G/K *are given by*

$$r_{(j,t)} = \check{r}_j - \frac{1}{2d_{j,t}} \sum_i \sum_{j',t'} \frac{z_i}{y_j y_{j'}} \left[\begin{matrix} i \\ (j,t) \ (j',t') \end{matrix} \right], \qquad (6)$$

where $\check{r}_j$ *are the components of Ricci tensor* $\check{r}$ *for the metric* $\check{g}$ *on* G/L.

Proof. Let $\{e_\alpha^{(j,t)}, e_\beta^{(i)}\}$ be a B-orthonormal basis adapted to the decomposition of $\mathfrak{p} \oplus \mathfrak{n} = \sum_j \sum_{t=1}^{k_j} \mathfrak{m}_{j,t} \oplus \sum_i \mathfrak{n}_i$ (with $e_\alpha^{(j,t)} \in \mathfrak{m}_{j,t}$ and $e_\beta^{(i)} \in \mathfrak{n}_i$). Put $X_\alpha^{(j,t)} = \frac{1}{\sqrt{y_j}} e_\alpha^{(j,t)}$ and $X_\beta^{(i)} = \frac{1}{\sqrt{z_i}} e_\beta^{(i)}$. Then $\left\{ X_\alpha^{(j,t)}, X_\beta^{(i)} \right\}$ is an orthonormal basis of $\mathfrak{p} \oplus \mathfrak{n}$ for the metric g. Then we have

$$\sum_{\gamma=1}^{d_{j,t}} r(X_\gamma^{(j,t)}, X_\gamma^{(j,t)}) = \sum_{\gamma=1}^{d_{j,t}} \check{r}(X_\gamma^{(j,t)}, X_\gamma^{(j,t)}) - \frac{1}{2} \sum_i \sum_{j',t'} \frac{z_i}{y_j y_{j'}} \left[\begin{matrix} i \\ (j,t) \ (j',t') \end{matrix} \right].$$

Noting that $\left\{ X_\gamma^{(j,t)} \right\}_{\gamma=1}^{d_{j,t}}$ is an orthonormal basis of $\mathfrak{m}_{j,t}$, we obtain our claim. $\qquad \square$

4. Decomposition associated to generalized flag manifolds

Let G be a compact semi-simple Lie group, $\mathfrak{g}$ the Lie algebra of G and $\mathfrak{h}$ a maximal abelian subalgebra of $\mathfrak{g}$. We denote by $\mathfrak{g}^\mathbb{C}$ and $\mathfrak{h}^\mathbb{C}$ the complexification of $\mathfrak{g}$ and $\mathfrak{h}$ respectively. We identify an element of the root system Δ of $\mathfrak{g}^\mathbb{C}$ relative to the Cartan subalgebra $\mathfrak{h}^\mathbb{C}$ with an element of $\mathfrak{h}_0 = \sqrt{-1}\mathfrak{h}$ by the duality defined by the Killing form of $\mathfrak{g}^\mathbb{C}$. Let $\Pi = \{\alpha_1, \cdots, \alpha_l\}$ be a fundamental system of Δ and $\{\Lambda_1, \cdots, \Lambda_l\}$ the fundamental weights of $\mathfrak{g}^{\mathbb{C}}$ corresponding to Π, that is

$$\frac{2(\Lambda_i, \alpha_j)}{(\alpha_j, \alpha_j)} = \delta_{ij} \qquad (1 \leq i, j \leq l).$$

Let Π_0 be a subset of Π and $\Pi - \Pi_0 = \{\alpha_{i_1}, \cdots, \alpha_{i_r}\}$, where $1 \leq i_1 < \cdots < i_r \leq l$. We put $[\Pi_0] = \Delta \cap \{\Pi_0\}_{\mathbb{Z}}$, where $\{\Pi_0\}_{\mathbb{Z}}$ denotes the subspace of $\mathfrak{h}_0$ generated by Π_0.

Consider the root space decomposition of $\mathfrak{g}^{\mathbb{C}}$ relative to $\mathfrak{h}^{\mathbb{C}}$:

$$\mathfrak{g}^{\mathbb{C}} = \mathfrak{h}^{\mathbb{C}} + \sum_{\alpha \in \Delta} \mathfrak{g}_{\alpha}^{\mathbb{C}}.$$

We define a parabolic subalgebra $\mathfrak{u}$ of $\mathfrak{g}^{\mathbb{C}}$ by

$$\mathfrak{u} = \mathfrak{h}^{\mathbb{C}} + \sum_{\alpha \in [\Pi_0] \cup \Delta^+} \mathfrak{g}_{\alpha}^{\mathbb{C}},$$

where Δ^+ is the set of all positive roots relative to Π.

Note that the nilradical $\mathfrak{n}$ of $\mathfrak{u}$ is given by

$$\mathfrak{n} = \sum_{\alpha \in \Delta^+ - [\Pi_0]} \mathfrak{g}_{\alpha}^{\mathbb{C}}.$$

We put $\Delta_{\mathfrak{m}}^+ = \Delta^+ - [\Pi_0]$.

Let $G^{\mathbb{C}}$ be a simply connected complex semi-simple Lie group whose Lie algebra is $\mathfrak{g}^{\mathbb{C}}$ and U the parabolic subgroup of $G^{\mathbb{C}}$ generated by $\mathfrak{u}$. Then the complex homogeneous manifold $G^{\mathbb{C}}/U$ is compact simply connected and G acts transitively on $G^{\mathbb{C}}/U$. Note also that $K = G \cap U$ is a connected closed subgroup of G, $G^{\mathbb{C}}/U = G/K$ as C^{∞}-manifolds, and $G^{\mathbb{C}}/U$ admits a G-invariant Kähler metric.

Let $\mathfrak{k}$ be the Lie algebra of K and $\mathfrak{k}^{\mathbb{C}}$ the complexification of $\mathfrak{k}$. Then we have a direct decomposition

$$\mathfrak{u} = \mathfrak{k}^{\mathbb{C}} \oplus \mathfrak{n}, \qquad \mathfrak{k}^{\mathbb{C}} = \mathfrak{h}^{\mathbb{C}} + \sum_{\alpha \in [\Pi_0]} \mathfrak{g}_{\alpha}^{\mathbb{C}}.$$

Take a Weyl basis $E_{-\alpha} \in \mathfrak{g}_{\alpha}^{\mathbb{C}}$ $(\alpha \in \Delta)$ with

$$[E_{\alpha}, E_{-\alpha}] = -\alpha \, (\alpha \in \Delta)$$
$$[E_{\alpha}, E_{\beta}] = \begin{cases} N_{\alpha, \beta} E_{\alpha+\beta} & \text{if} \quad \alpha + \beta \in \Delta \\ 0 & \text{if} \quad \alpha + \beta \notin \Delta, \end{cases}$$

where $N_{\alpha,\beta} = N_{-\alpha,-\beta} \in \mathbb{R}$. Then we have

$$\mathfrak{g} = \mathfrak{h} + \sum_{\alpha \in \Delta} \left\{ \mathbb{R}(E_\alpha + E_{-\alpha}) + \mathbb{R}\sqrt{-1}(E_\alpha - E_{-\alpha}) \right\}$$

and the Lie subalgebra $\mathfrak{k}$ is given by

$$\mathfrak{k} = \mathfrak{h} + \sum_{\alpha \in [\Pi_0]} \left\{ \mathbb{R}(E_\alpha + E_{-\alpha}) + \mathbb{R}\sqrt{-1}(E_\alpha - E_{-\alpha}) \right\}.$$

For integers $j_1, \cdots, j_r$ with $(j_1, \cdots, j_r) \neq (0, \cdots, 0)$, we put

$$\Delta(j_1, \cdots, j_r) = \left\{ \sum_{j=1}^{l} m_j \alpha_j \in \Delta^+ \;\middle|\; m_{i_1} = j_1, \cdots, m_{i_r} = j_r \right\}.$$

Note that $\Delta_{\mathfrak{m}}^+ = \Delta^+ - [\Pi_0] = \bigcup_{j_1,\cdots,j_r} \Delta(j_1, \cdots, j_r)$.

For $\Delta(j_1, \cdots, j_r) \neq \emptyset$, we define an $\mathrm{Ad}_G(K)$-invariant subspace $\mathfrak{m}(j_1, \cdots, j_r)$ of $\mathfrak{g}$ by

$$\mathfrak{m}(j_1, \cdots, j_r) = \sum_{\alpha \in \Delta(j_1,\cdots,j_r)} \left\{ \mathbb{R}(E_\alpha + E_{-\alpha}) + \mathbb{R}\sqrt{-1}(E_\alpha - E_{-\alpha}) \right\}.$$

Then we have a decomposition of $\mathfrak{m}$ into mutually non-equivalent irreducible $\mathrm{Ad}_G(K)$-modules $\mathfrak{m}(j_1, \cdots, j_r)$:

$$\mathfrak{m} = \sum_{j_1,\cdots,j_r} \mathfrak{m}(j_1, \cdots, j_r).$$

We put $\mathfrak{t} = \left\{ H \in \mathfrak{h}_0 \mid (H, \Pi_0) = 0 \right\}$. Then $\{\Lambda_{i_1}, \cdots, \Lambda_{i_r}\}$ is a basis of $\mathfrak{t}$. Put $\mathfrak{s} = \sqrt{-1}\mathfrak{t}$. Then the Lie algebra $\mathfrak{k}$ is given by $\mathfrak{k} = \mathfrak{z}(\mathfrak{s})$ (the Lie algebra of centralizer of a torus S in G).

We consider the restriction map

$$\kappa : \mathfrak{h}_0^* \to \mathfrak{t}^* \qquad \alpha \mapsto \alpha|_\mathfrak{t}$$

and set $\Delta_T = \kappa(\Delta)$. The elements of Δ_T are called T-roots.

There exists a 1-1 correspondence between T-roots ξ and irreducible submodules $\mathfrak{m}_\xi$ of the $\mathrm{Ad}_G(K)$-module $\mathfrak{m}^\mathbb{C}$ given by

$$\Delta_T \ni \xi \mapsto \mathfrak{m}_\xi = \sum_{\kappa(\alpha)=\xi} \mathfrak{g}_\alpha^\mathbb{C}.$$

Thus we have a decomposition of the $\mathrm{Ad}_G(K)$-module $\mathfrak{m}^\mathbb{C}$:

$$\mathfrak{m}^\mathbb{C} = \sum_{\xi \in \Delta_T} \mathfrak{m}_\xi.$$

Denote by Δ_T^+ the set of all positive T-roots, that is, the restricton of the system Δ^+. Then we have $\mathfrak{n} = \sum\limits_{\xi \in \Delta_T^+} \mathfrak{m}_\xi$. Denote by τ the complex conjugation of $\mathfrak{g}^\mathbb{C}$ with respect to $\mathfrak{g}$ (note that τ interchanges $\mathfrak{g}_\alpha^\mathbb{C}$ and $\mathfrak{g}_{-\alpha}^\mathbb{C}$) and by $\mathfrak{v}^\tau$ the set of fixed points of τ in a complex vector subspace $\mathfrak{v}$ of $\mathfrak{g}^\mathbb{C}$. Thus we have a decomposition of $\mathrm{Ad}_G(K)$-module $\mathfrak{m}$ into irreducible submodules:

$$\mathfrak{m} = \sum_{\xi \in \Delta_T^+} (\mathfrak{m}_\xi + \mathfrak{m}_{-\xi})^\tau.$$

There exists a natural 1-1 correspondence between Δ_T^+ and the set $\{\Delta(j_1, \cdots, j_r) \neq \emptyset\}$. For a generalized flag manifold G/K, we have a decomposition of $\mathfrak{m}$ into mutually non-equivalent irreducible $\mathrm{Ad}_G(H)$-modules:

$$\mathfrak{m} = \sum_{\xi \in \Delta_T^+} (\mathfrak{m}_\xi + \mathfrak{m}_{-\xi})^\tau = \sum_{j_1, \cdots, j_r} \mathfrak{m}(j_1, \cdots, j_r).$$

Thus a G-invariant metric g on G/K can be written as

$$g = \sum_{\xi \in \Delta_T^+} x_\xi B|_{(\mathfrak{m}_\xi + \mathfrak{m}_{-\xi})^\tau} = \sum_{j_1, \cdots, j_r} x_{j_1 \cdots j_r} B|_{\mathfrak{m}(j_1, \cdots, j_r)} \tag{7}$$

for positive real numbers x_ξ, $x_{j_1 \cdots j_r}$.

Put $Z_\mathfrak{t} = \left\{ \Lambda \in \mathfrak{t} \;\middle|\; \dfrac{2(\Lambda, \alpha)}{(\alpha, \alpha)} \in \mathbb{Z} \text{ for each } \alpha \in \Delta \right\}$. Then $Z_\mathfrak{t}$ is a lattice of $\mathfrak{t}$ generated by $\{\Lambda_{i_1}, \cdots, \Lambda_{i_r}\}$. For each $\Lambda \in Z_\mathfrak{t}$ there exists a unique holomorphic character χ_Λ of U such that $\chi_\Lambda(\exp H) = \exp \Lambda(H)$ for each $H \in \mathfrak{h}^\mathbb{C}$. Then the correspondence $\Lambda \to \chi_\Lambda$ gives an isomorphism of $Z_\mathfrak{t}$ to the group of holomorphic characters of U.

Let F_Λ denote the holomorphic line bundle on $G^\mathbb{C}/U$ associated to the principal bundle $U \to G^\mathbb{C} \to G^\mathbb{C}/U$ by the holomorphic character χ_Λ, and $H(G^\mathbb{C}/U, \mathbb{C}^*)$ the group of isomorphism classes of holomorphic line bundles on $G^\mathbb{C}/U$.

The correspondence $\Lambda \mapsto F_\Lambda : Z_\mathfrak{t} \to H(G^\mathbb{C}/U, \mathbb{C}^*)$ induces a homomorphism. Also the correspondence $F \mapsto c_1(F)$ defines a homomorphism of $H(G^\mathbb{C}/U, \mathbb{C}^*)$ to $H^2(M, \mathbb{Z})$.

Then it is known that homomorphisms

$$Z_\mathfrak{t} \xrightarrow{F} H(G^\mathbb{C}/U, \mathbb{C}^*) \xrightarrow{c_1} H^2(M, \mathbb{Z})$$

are in fact isomorphisms. In particular, 2nd Betti number $b_2(M)$ of M is given by

$$b_2(M) = \dim \mathfrak{t} = \text{ the cardinality of } \Pi - \Pi_0.$$

5. Kähler Einstein metric of a generalized flag manifold

Put $Z_t^+ = \{\lambda \in Z_t \mid (\lambda, \alpha) > 0 \text{ for } \alpha \in \Pi - \Pi_0\}$. Then we have $Z_t^+ = \sum_{\alpha \in \Pi - \Pi_0} \mathbb{Z}^+ \Lambda_\alpha$. We define an element $\delta_m \in \sqrt{-1}\mathfrak{h}$ by $\delta_m = \frac{1}{2} \sum_{\alpha \in \Delta_m^+} \alpha$.

Let $c_1(M)$ be the first Chern class of M. Then we have

$$2\delta_m \in Z_t^+, \quad c_1(M) = c_1(F_{2\delta_m}).$$

Put $k_\alpha = \dfrac{2(2\delta_m, \alpha)}{(\alpha, \alpha)}$ for $\alpha \in \Pi - \Pi_0$. Then $2\delta_m = \sum_{\alpha \in \Pi - \Pi_0} k_\alpha \Lambda_\alpha = k_{\alpha_{i_1}} \Lambda_{\alpha_{i_1}} + \cdots + k_{\alpha_{i_r}} \Lambda_{\alpha_{i_r}}$ and each $k_{\alpha_{i_s}}$ is a positive integer.

The G-invariant metric $g_{2\delta_m}$ on G/K corresponding to $2\delta_m$, which is a Kähler Einstein metric, is given by

$$g_{2\delta_m} = \sum_{j_1, \cdots, j_r} \left(\sum_{\ell=1}^r k_{\alpha_{i_\ell}} j_\ell \frac{(\alpha_{j_\ell}, \alpha_{j_\ell})}{2} \right) B|_{\mathfrak{m}(j_1, \cdots, j_r)}.$$

Example 5.1. For generalized flag manifolds $G/K = Sp(n)/(U(p) \times U(q) \times Sp(n - p - q))$ where $n \geq 3$, $p, q \geq 1$, we see that $\Pi - \Pi_0 = \{\alpha_p, \alpha_{p+q}\}$ and $2\delta_m = (p + q)\Lambda_{\alpha_p} + (2n - 2p - q + 1)\Lambda_{\alpha_{p+q}}$. Thus the Kähler Einstein metric $g_{2\delta_m}$ on G/K is given by

$$
\begin{aligned}
g_{2\delta_m} = {}& (p + q)B|_{\mathfrak{m}(1,0)} + (2n - 2p - q + 1)B|_{\mathfrak{m}(0,1)} \\
& + (2n - p + 1)B|_{\mathfrak{m}(1,1)} + 2(2n - 2p - q + 1)B|_{\mathfrak{m}(0,2)} \\
& + (4n - 3p - q + 2)B|_{\mathfrak{m}(1,2)} + 2(2n - p + 1)B|_{\mathfrak{m}(2,2)}.
\end{aligned}
\tag{8}
$$

6. Einstein metrics on generalized flag manifolds with two isotropy summands

From now on we assume that the Lie group G is simple. For a generalized flag manifold G/K, we denote by q the number of mutually non-equivalent irreducible $\mathrm{Ad}_G(K)$-modules $\mathfrak{m}(j_1, \cdots, j_r)$ with $\mathfrak{m} = \sum_{j_1, \cdots, j_r} \mathfrak{m}(j_1, \cdots, j_r)$.

In the case when $q = 1$, it is known that G/K is an irreducible Hermitian symmetric space with the symmetric pair $(\mathfrak{g}, \mathfrak{k})$. In the case when $q = 2$, we have two G-invariant Einstein metrics on G/K. One is Kähler Einstein metric and the other is non-Kähler Einstein metric. In fact, we see that the case $q = 2$ occurs only in the case $r = b_2(G/K) = 1$ and $\mathfrak{m} = \mathfrak{m}(1) \oplus \mathfrak{m}(2)$ (cf. [6]). Note that only $\begin{bmatrix} 2 \\ 11 \end{bmatrix}$ is non-zero. Put $d_1 = \dim \mathfrak{m}(1)$ and $d_2 = \dim \mathfrak{m}(2)$.

For a G-invariant metric $\langle \ , \ \rangle = x_1 \cdot B|_{\mathfrak{m}(1)} + x_2 \cdot B|_{\mathfrak{m}(2)}$, the components r_1, r_2 of Ricci tensor r of the metric $\langle \ , \ \rangle$ are given by

$$
\begin{cases}
r_1 = \dfrac{1}{2x_1} - \dfrac{x_2}{2\,d_1\,x_1{}^2} \begin{bmatrix} 2 \\ 11 \end{bmatrix} \\[4mm]
r_2 = \dfrac{1}{2x_2} - \dfrac{1}{2\,d_2\,x_2} \begin{bmatrix} 1 \\ 21 \end{bmatrix} + \dfrac{x_2}{4\,d_2\,x_1{}^2} \begin{bmatrix} 2 \\ 11 \end{bmatrix}.
\end{cases}
\tag{9}
$$

Since the metric $(\ , \ ) = 1 \cdot B|_{\mathfrak{m}(1)} + 2 \cdot B|_{\mathfrak{m}(2)}$ is Kähler Einstein, we see that $\begin{bmatrix} 2 \\ 11 \end{bmatrix} = \dfrac{d_1 d_2}{d_1 + 4d_2}$. Note that a G-invariant metric $\langle \ , \ \rangle = x_1 \cdot B|_{\mathfrak{m}(1)} + x_2 \cdot B|_{\mathfrak{m}(2)}$ is Einstein if and only if $r_1 = r_2$. We normalize the equation $r_1 = r_2$ by putting $x_1 = 1$. Then we see that the equation $r_1 = r_2$ is reduced to a quadratic equation of x_2 and we have solutions $x_2 = 2$ and $x_2 = \dfrac{4d_2}{d_1 + 2d_2}$. Since $x_2 = \dfrac{4d_2}{d_1 + 2d_2} \neq 2$, the Einstein metric $1 \cdot B|_{\mathfrak{m}(1)} + \dfrac{4d_2}{d_1 + 2d_2} \cdot B|_{\mathfrak{m}(2)}$ is non-Kähler.

Example 6.1. For generalized flag manifolds $Sp(n)/(U(a) \times Sp(n - a))$ where $1 \leq a \leq n-1$, we see that $d_1 = 4a(n-a)$ and $d_2 = a(a+1)$. Thus we have $\begin{bmatrix} 2 \\ 11 \end{bmatrix} = \dfrac{d_1 d_2}{d_1 + 4d_2} = \dfrac{a(a+1)(n-a)}{n+1}$ and non-Kähler Einstein metric is given by $x_1 = 1$ and $x_2 = \dfrac{2(a+1)}{2n - a + 1}$.

7. Generalized flag manifolds of type $Sp(n)$

We now consider the case $G/K = Sp(n)/(U(p) \times U(q) \times Sp(n - p - q))$ where $n \geq 3$, $p, q \geq 1$ and $2 \leq p + q \leq n - 1$. Its corresponding painted Dynkin diagram is given by

$$
\begin{array}{ccccccccc}
\alpha_1 & & \alpha_{p-1}\ \alpha_p & & & \alpha_{p+q} & & \alpha_{n-1}\ \alpha_n \\
\circ\!\!-\!\!\cdots\!\!-\!\!\circ\!\!-\!\!\bullet\!\!-\!\!\circ\!\!-\!\!\cdots\!\!-\!\!\bullet\!\!-\!\!\cdots\!\!-\!\!\times\!\!\Longrightarrow\!\!\circ \\
2 \qquad\quad 2 \quad 2 \quad 2 \qquad\quad 2 \qquad\quad 2 \qquad 1
\end{array}.
$$

The set of all positive T-roots Δ_T^+ is given by

$$
\Delta_T^+ = \{\xi_1, \ \xi_2, \ \xi_1 + \xi_2, \ 2\xi_2, \ \xi_1 + 2\xi_2, \ 2\xi_1 + 2\xi_2\}
$$

where $\xi_1 = \kappa(\alpha_p)$ and $\xi_2 = \kappa(\alpha_{p+q})$ and we have a decomposition of $\mathfrak{m}$ into 6 mutually non-equivalent irreducible $\mathrm{Ad}_G(H)$-modules:

$$
\mathfrak{m} = \mathfrak{m}(1,0) \oplus \mathfrak{m}(0,1) \oplus \mathfrak{m}(1,1) \oplus \mathfrak{m}(0,2) \oplus \mathfrak{m}(1,2) \oplus \mathfrak{m}(2,2).
$$

We put $\mathfrak{m}_1 = \mathfrak{m}(1,0)$, $\mathfrak{m}_2 = \mathfrak{m}(0,1)$, $\mathfrak{m}_3 = \mathfrak{m}(1,1)$, $\mathfrak{m}_4 = \mathfrak{m}(0,2)$, $\mathfrak{m}_5 = \mathfrak{m}(1,2)$ and $\mathfrak{m}_6 = \mathfrak{m}(2,2)$. Then we see that $\dim \mathfrak{m}_1 = 2pq$, $\dim \mathfrak{m}_2 = 4q(n - p - q)$, $\dim \mathfrak{m}_3 = 4p(n - p - q)$, $\dim \mathfrak{m}_4 = q(q + 1)$, $\dim \mathfrak{m}_5 = 2pq$ and $\dim \mathfrak{m}_6 = p(p + 1)$.

It is easy to see that $\begin{bmatrix} 3 \\ 12 \end{bmatrix}$, $\begin{bmatrix} 5 \\ 14 \end{bmatrix}$, $\begin{bmatrix} 6 \\ 15 \end{bmatrix}$, $\begin{bmatrix} 4 \\ 22 \end{bmatrix}$, $\begin{bmatrix} 5 \\ 23 \end{bmatrix}$ and $\begin{bmatrix} 6 \\ 33 \end{bmatrix}$ are non-zero. We write G-invariant metrics g on G/K as

$$g = x_1 B|_{\mathfrak{m}_1} + x_2 B|_{\mathfrak{m}_2} + x_3 B|_{\mathfrak{m}_3} + x_4 B|_{\mathfrak{m}_4} + x_5 B|_{\mathfrak{m}_5} + x_6 B|_{\mathfrak{m}_6} \quad (10)$$

where x_j $(j = 1, \cdots, 6)$ are positive numbers.

Now from Lemma 2.1, we obtain the following proposition.

Proposition 7.1. *The components r_i $(i = 1, \cdots, 6)$ of the Ricci tensor for a G-invariant Riemannian metric on G/K determined by (10) are given as follows:*

$$r_1 = \frac{1}{2x_1} + \frac{1}{2d_1}\begin{bmatrix} 3 \\ 12 \end{bmatrix}\left(\frac{x_1}{x_2 x_3} - \frac{x_2}{x_1 x_3} - \frac{x_3}{x_1 x_2}\right)$$
$$+ \frac{1}{2d_1}\begin{bmatrix} 5 \\ 14 \end{bmatrix}\left(\frac{x_1}{x_4 x_5} - \frac{x_5}{x_1 x_4} - \frac{x_4}{x_1 x_5}\right) + \frac{1}{2d_1}\begin{bmatrix} 6 \\ 15 \end{bmatrix}\left(\frac{x_1}{x_5 x_6} - \frac{x_6}{x_1 x_5} - \frac{x_5}{x_1 x_6}\right),$$

$$r_2 = \frac{1}{2x_2} + \frac{1}{2d_2}\begin{bmatrix} 3 \\ 12 \end{bmatrix}\left(\frac{x_2}{x_1 x_3} - \frac{x_1}{x_2 x_3} - \frac{x_3}{x_1 x_2}\right)$$
$$- \frac{1}{2d_2}\begin{bmatrix} 4 \\ 22 \end{bmatrix}\frac{x_4}{x_2^2} + \frac{1}{2d_2}\begin{bmatrix} 5 \\ 23 \end{bmatrix}\left(\frac{x_2}{x_3 x_5} - \frac{x_5}{x_2 x_3} - \frac{x_3}{x_2 x_5}\right),$$

$$r_3 = \frac{1}{2x_3} + \frac{1}{2d_3}\begin{bmatrix} 3 \\ 12 \end{bmatrix}\left(\frac{x_3}{x_1 x_2} - \frac{x_2}{x_1 x_3} - \frac{x_1}{x_2 x_3}\right)$$
$$+ \frac{1}{2d_3}\begin{bmatrix} 5 \\ 23 \end{bmatrix}\left(\frac{x_3}{x_2 x_5} - \frac{x_5}{x_2 x_3} - \frac{x_2}{x_3 x_5}\right) - \frac{1}{2d_3}\begin{bmatrix} 6 \\ 33 \end{bmatrix}\frac{x_6}{x_3^2},$$

$$r_4 = \frac{1}{2x_4} + \frac{1}{2d_4}\begin{bmatrix} 5 \\ 14 \end{bmatrix}\left(\frac{x_4}{x_1 x_5} - \frac{x_5}{x_1 x_4} - \frac{x_1}{x_4 x_5}\right) + \frac{1}{4d_4}\begin{bmatrix} 4 \\ 22 \end{bmatrix}\left(-\frac{2}{x_4} + \frac{x_4}{x_2^2}\right),$$

$$r_5 = \frac{1}{2x_5} + \frac{1}{2d_5}\begin{bmatrix} 5 \\ 23 \end{bmatrix}\left(\frac{x_5}{x_2 x_3} - \frac{x_2}{x_3 x_5} - \frac{x_3}{x_2 x_5}\right)$$
$$+ \frac{1}{2d_5}\begin{bmatrix} 5 \\ 14 \end{bmatrix}\left(\frac{x_5}{x_1 x_4} - \frac{x_1}{x_4 x_5} - \frac{x_4}{x_1 x_5}\right) + \frac{1}{2d_5}\begin{bmatrix} 6 \\ 15 \end{bmatrix}\left(\frac{x_5}{x_1 x_6} - \frac{x_6}{x_1 x_5} - \frac{x_1}{x_5 x_6}\right),$$

$$r_6 = \frac{1}{2x_6} + \frac{1}{2d_6}\begin{bmatrix} 6 \\ 15 \end{bmatrix}\left(\frac{x_6}{x_1 x_5} - \frac{x_5}{x_1 x_6} - \frac{x_1}{x_6 x_5}\right) + \frac{1}{4d_6}\begin{bmatrix} 6 \\ 33 \end{bmatrix}\left(-\frac{2}{x_6} + \frac{x_6}{x_3^2}\right).$$

Now we compute the non-negative numbers $\begin{bmatrix} k \\ ij \end{bmatrix}$.

Proposition 7.2.

$$\begin{bmatrix} 3 \\ 12 \end{bmatrix} = \frac{pq(n-p-q)}{n+1}, \quad \begin{bmatrix} 5 \\ 14 \end{bmatrix} = \frac{pq(q+1)}{2(n+1)},$$

$$\begin{bmatrix} 6 \\ 15 \end{bmatrix} = \frac{p(p+1)q}{2(n+1)}, \quad \begin{bmatrix} 4 \\ 22 \end{bmatrix} = \frac{q(q+1)(n-p-q)}{n+1},$$

$$\begin{bmatrix} 5 \\ 23 \end{bmatrix} = \frac{pq(n-p-q)}{n+1}, \quad \begin{bmatrix} 6 \\ 33 \end{bmatrix} = \frac{p(p+1)(n-p-q)}{n+1}.$$

Proof. We consider the subgroups $L_1 = U(p) \times Sp(n-p)$ and $L_2 = U(p+q) \times Sp(n-p-q)$. Then we have Riemannian submersions $G/K \to G/L_1$ and $G/K \to G/L_2$. Note that G/L_1 and G/L_2 are generalized flag manifolds with two isotropy summands which are considered in Example 6.1.

We consider a G-invariant metric g_1 on G/K defined by a Riemannian submersion $\pi : (G/K, g_1) \to (G/L_1, \check{g}_1)$ with totally geodesic fibers isometric to $(L_1/K, \hat{g}_1)$. Since L_1/K is isomorphic to $Sp(n-p)/(U(q) \times Sp(n-p-q))$, we see that the metric g_1 on G/K and the metric $\check{g}_1$ on G/L_1 are of the forms

$$g_1 = y_1 B|_{\mathfrak{p}_1} + y_2 B|_{\mathfrak{p}_2} + z_1 B|_{\mathfrak{n}_1} + z_2 B|_{\mathfrak{n}_2} \quad \text{and} \quad \check{g}_1 = y_1 B|_{\mathfrak{p}_1} + y_2 B|_{\mathfrak{p}_2}.$$

We decompose irreducible components $\mathfrak{p}_1$ and $\mathfrak{p}_2$ into irreducible $\mathrm{Ad}(K)$-modules. Then we obtain that

$$\mathfrak{p}_1 = \mathfrak{m}_1 \oplus \mathfrak{m}_3 \oplus \mathfrak{m}_5, \quad \mathfrak{p}_2 = \mathfrak{m}_6, \quad \mathfrak{n}_1 = \mathfrak{m}_2, \quad \mathfrak{n}_2 = \mathfrak{m}_4.$$

Thus the metric g_1 can be written as

$$g_1 = y_1 B|_{\mathfrak{m}_1} + z_1 B|_{\mathfrak{m}_2} + y_1 B|_{\mathfrak{m}_3} + z_2 B|_{\mathfrak{m}_4} + y_1 B|_{\mathfrak{m}_5} + y_2 B|_{\mathfrak{m}_6}. \quad (11)$$

Since the metric (11) is a special case of the metric (10), from Proposition 7.1 we see that the components r_1, r_3 of the Ricci tensor for a metric g_1 on G/K are given by

$$r_1 = \frac{1}{2y_1} - \frac{1}{2d_1} \begin{bmatrix} 3 \\ 12 \end{bmatrix} \frac{z_1}{y_1{}^2} - \frac{1}{2d_1} \begin{bmatrix} 5 \\ 14 \end{bmatrix} \frac{z_2}{y_1{}^2} - \frac{1}{2d_1} \begin{bmatrix} 6 \\ 15 \end{bmatrix} \frac{y_2}{y_1{}^2}, \quad (12)$$

$$r_3 = \frac{1}{2y_1} - \frac{1}{2d_3} \begin{bmatrix} 3 \\ 12 \end{bmatrix} \frac{z_1}{y_1{}^2} - \frac{1}{2d_3} \begin{bmatrix} 5 \\ 23 \end{bmatrix} \frac{z_1}{y_1{}^2} - \frac{1}{2d_3} \begin{bmatrix} 6 \\ 33 \end{bmatrix} \frac{y_2}{y_1{}^2}. \quad (13)$$

From Example 6.1, we see that the component $\check{r}_1$ of the Ricci tensor for a metric $\check{g}_1$ on G/L_1 is given by

$$\check{r}_1 = \frac{1}{2y_1} - \frac{y_2}{2\widetilde{d}_1 y_1{}^2} \frac{\widetilde{d}_1 \widetilde{d}_2}{\widetilde{d}_1 + 4\widetilde{d}_2} \quad (14)$$

where $\widetilde{d_1} = \dim \mathfrak{p}_1 = 4p(n-p)$ and $\widetilde{d_2} = \dim \mathfrak{p}_2 = p(p+1)$.

From Lemma 3.1, (12), (13) and (14), we obtain that $\begin{bmatrix} 6 \\ 15 \end{bmatrix} = \dfrac{p(p+1)q}{2(n+1)}$ and $\begin{bmatrix} 6 \\ 33 \end{bmatrix} = \dfrac{p(p+1)(n-p-q)}{n+1}$.

We also consider a G-invariant metric g_2 on G/K defined by a Riemannian submersion $\pi : (G/K, g_2) \to (G/L_2, \check{g}_2)$ with totally geodesic fibers isometric to $(L_2/K, \hat{g}_2)$. Since L_2/K is isomorphic to $SU(p+q)/S(U(q) \times U(q))$, we see that the metric g_2 on G/K and the metric $\check{g}_2$ on G/L_2 are of the forms

$$g_2 = v_1 B|_{\mathfrak{q}_1} + v_2 B|_{\mathfrak{q}_2} + w_1 B|_{\mathfrak{l}_1} \quad \text{and} \quad \check{g}_2 = v_1 B|_{\mathfrak{q}_1} + v_2 B|_{\mathfrak{q}_2}.$$

We decompose irreducible components $\mathfrak{q}_1$ and $\mathfrak{q}_2$ into irreducible $\mathrm{Ad}(K)$-modules. Then we obtain that

$$\mathfrak{q}_1 = \mathfrak{m}_2 \oplus \mathfrak{m}_3, \quad \mathfrak{q}_2 = \mathfrak{m}_4 \oplus \mathfrak{m}_5 \oplus \mathfrak{m}_6, \quad \mathfrak{l}_1 = \mathfrak{m}_1.$$

Thus the metric g_2 can be written as

$$g_2 = w_1 B|_{\mathfrak{m}_1} + v_1 B|_{\mathfrak{m}_2} + v_1 B|_{\mathfrak{m}_3} + v_2 B|_{\mathfrak{m}_4} + v_2 B|_{\mathfrak{m}_5} + v_2 B|_{\mathfrak{m}_6}. \quad (15)$$

Since the metric (15) is a special case of the metric (10), from Proposition 7.1 we see that the components r_4, r_5 of the Ricci tensor for a G-invariant Riemannian metric g_2 on G/K are given by

$$r_4 = \frac{1}{2v_2} - \frac{1}{2d_4}\begin{bmatrix} 5 \\ 14 \end{bmatrix}\frac{w_1}{v_2{}^2} + \frac{1}{4d_4}\begin{bmatrix} 4 \\ 22 \end{bmatrix}\left(-\frac{2}{v_2} + \frac{v_2}{v_1{}^2}\right), \quad (16)$$

$$r_5 = \frac{1}{2v_2} + \frac{1}{2d_5}\begin{bmatrix} 5 \\ 23 \end{bmatrix}\left(-\frac{2}{v_2} + \frac{v_2}{v_1{}^2}\right) - \frac{1}{2d_5}\left(\begin{bmatrix} 5 \\ 14 \end{bmatrix} + \begin{bmatrix} 6 \\ 15 \end{bmatrix}\right)\frac{w_1}{v_2{}^2}. \quad (17)$$

From Example 6.1, we see that the component $\check{r}_2$ of the Ricci tensor for a metric $\check{g}_2$ on G/L_2 is given by

$$\check{r}_2 = \frac{1}{2y_2} - \frac{1}{2\widetilde{d_2}\,y_2}\frac{\widetilde{d_1}\widetilde{d_2}}{\widetilde{d_1}+4\widetilde{d_2}} + \frac{y_2}{4\widetilde{d_2}\,y_1{}^2}\frac{\widetilde{d_1}\widetilde{d_2}}{\widetilde{d_1}+4\widetilde{d_2}}, \quad (18)$$

where $\widetilde{d_1} = \dim \mathfrak{q}_1 = 4(p+q)(n-p-q)$ and $\widetilde{d_2} = \dim \mathfrak{q}_2 = (p+q)(p+q+1)$.

From Lemma 3.1, (16), (17) and (18), we obtain that $\begin{bmatrix} 4 \\ 22 \end{bmatrix} = \dfrac{q(q+1)(n-p-q)}{n+1}$ and $\begin{bmatrix} 5 \\ 23 \end{bmatrix} = \dfrac{pq(n-p-q)}{n+1}$.

Now taking into account the explicit form (8) of the Kähler Einstein metric $g_{2\delta_{\mathfrak{m}}}$ on G/K in Example 5.1, and substituting these values for $x_1, x_2, x_3, x_4, x_5, x_6$ in the components r_i of the Ricci tensor, we can find the values of $\begin{bmatrix} 3 \\ 12 \end{bmatrix}$ and $\begin{bmatrix} 5 \\ 14 \end{bmatrix}$. In fact, substituting also the values for

$\begin{bmatrix} 6 \\ 15 \end{bmatrix}$, $\begin{bmatrix} 6 \\ 33 \end{bmatrix}$, $\begin{bmatrix} 4 \\ 22 \end{bmatrix}$ and $\begin{bmatrix} 5 \\ 23 \end{bmatrix}$ into $r_2 = r_3$ and $r_4 = r_5$, we obtain that

$\begin{bmatrix} 3 \\ 12 \end{bmatrix} = \dfrac{pq(n-p-q)}{n+1}$ and $\begin{bmatrix} 5 \\ 14 \end{bmatrix} = \dfrac{pq(q+1)}{2(n+1)}$. $\qquad\qquad \square$

8. Proof of theorems

We normalize the system of equations as

$$r_1 = 1, \quad r_1 = r_5, \quad r_2 = r_3, \quad r_3 = r_4, \quad r_4 = r_5, \quad r_5 = r_6. \qquad (19)$$

8.1. Case of $Sp(3)/(U(1) \times U(1) \times Sp(1))$

For $Sp(3)/(U(1) \times U(1) \times Sp(1))$, the system of equations (19) reduces to the following system of polynomial equations:

$$\left\{ \begin{aligned}
& x_1{}^2 x_2 x_3 x_4 + x_1{}^2 x_2 x_3 x_6 + x_1{}^2 x_4 x_5 x_6 - 16 x_1 x_2 x_3 x_4 x_5 x_6 \\
& - x_2{}^2 x_4 x_5 x_6 - x_2 x_3 x_4{}^2 x_6 - x_2 x_3 x_4 x_5{}^2 + 8 x_2 x_3 x_4 x_5 x_6 \\
& - x_2 x_3 x_4 x_6{}^2 - x_2 x_3 x_5{}^2 x_6 - x_3{}^2 x_4 x_5 x_6 = 0, \\
& (x_1 - x_5)(2x_1 x_2 x_3 x_4 + 2x_1 x_2 x_3 x_6 + x_1 x_4 x_5 x_6 + x_2{}^2 x_4 x_6 \\
& + 2x_2 x_3 x_4 x_5 - 8x_2 x_3 x_4 x_6 + 2x_2 x_3 x_5 x_6 + x_3{}^2 x_4 x_6) = 0, \\
& x_1 x_2{}^3 x_3 - 8x_1 x_2{}^2 x_3 x_5 + x_1 x_2{}^2 x_5 x_6 - x_1 x_2 x_3{}^3 + 8x_1 x_2 x_3{}^2 x_5 \\
& - x_1 x_3{}^2 x_4 x_5 + x_2{}^3 x_3 x_5 - x_2 x_3{}^3 x_5 = 0, \\
& 2x_1{}^2 x_2{}^2 x_3{}^2 - x_1{}^2 x_2 x_3 x_4 x_5 - x_1 x_2{}^3 x_3 x_4 - 12 x_1 x_2{}^2 x_3{}^2 x_5 \\
& + 16 x_1 x_2{}^2 x_3 x_4 x_5 - 2x_1 x_2{}^2 x_4 x_5 x_6 + x_1 x_2 x_3{}^3 x_4 - x_1 x_2 x_3 x_4 x_5{}^2 \\
& - 2x_1 x_3{}^2 x_4{}^2 x_5 - x_2{}^3 x_3 x_4 x_5 - 2x_2{}^2 x_3{}^2 x_4{}^2 + 2x_2{}^2 x_3{}^2 x_5{}^2 \\
& + x_2 x_3{}^3 x_4 x_5 = 0, \\
& x_1{}^2 x_2{}^2 x_3 x_4 + x_1 x_2{}^3 x_4 x_6 - 8x_1 x_2{}^2 x_3 x_4 x_6 + 6x_1 x_2{}^2 x_3 x_5 x_6 \\
& + x_1 x_2 x_3{}^2 x_4 x_6 - x_1 x_2 x_4 x_5{}^2 x_6 + x_1 x_3 x_4{}^2 x_5 x_6 + 2x_2{}^2 x_3 x_4{}^2 x_6 \\
& - x_2{}^2 x_3 x_4 x_5{}^2 + x_2{}^2 x_3 x_4 x_6{}^2 - 2x_2{}^2 x_3 x_5{}^2 x_6 = 0, \\
& - x_1{}^2 x_2 x_3{}^2 x_6 - x_1 x_2{}^2 x_3 x_4 x_6 - 6x_1 x_2 x_3{}^2 x_4 x_5 + 8x_1 x_2 x_3{}^2 x_4 x_6 \\
& - x_1 x_2 x_4 x_5 x_6{}^2 - x_1 x_3{}^3 x_4 x_6 + x_1 x_3 x_4 x_5{}^2 x_6 - x_2 x_3{}^2 x_4{}^2 x_6 \\
& + 2x_2 x_3{}^2 x_4 x_5{}^2 - 2x_2 x_3{}^2 x_4 x_6{}^2 + x_2 x_3{}^2 x_5{}^2 x_6 = 0.
\end{aligned} \right. \qquad (20)$$

Case of $x_5 \neq x_1$. We first consider the case when $x_3 = x_2$. Then we see that to find non-zero solutions of the system of polynomial equations (20) reduces to the following system of polynomial equations:

$$\begin{cases} x_1{}^2x_2{}^2x_4 + x_1{}^2x_2{}^2x_6 + x_1{}^2x_4x_5x_6 - 16x_1x_2{}^2x_4x_5x_6 - x_2{}^2x_4{}^2x_6 \\ \quad -x_2{}^2x_4x_5{}^2 + 6x_2{}^2x_4x_5x_6 - x_2{}^2x_4x_6{}^2 - x_2{}^2x_5{}^2x_6 = 0, \\ 2x_1x_2{}^2x_4 + 2x_1x_2{}^2x_6 + x_1x_4x_5x_6 + 2x_2{}^2x_4x_5 \\ \quad -6x_2{}^2x_4x_6 + 2x_2{}^2x_5x_6 = 0, \\ x_4 - x_6 = 0, \\ 2x_1{}^2x_2{}^2 - x_1{}^2x_4x_5 - 12x_1x_2{}^2x_5 + 16x_1x_2x_4x_5 - 2x_1x_4{}^2x_5 \\ \quad -x_1x_4x_5{}^2 - 2x_1x_4x_5x_6 - 2x_2{}^2x_4{}^2 + 2x_2{}^2x_5{}^2 = 0, \\ x_1{}^2x_2{}^2x_4 - 6x_1x_2{}^2x_4x_6 + 6x_1x_2{}^2x_5x_6 + x_1x_4{}^2x_5x_6 - x_1x_4x_5{}^2x_6 \\ \quad +2x_2{}^2x_4{}^2x_6 - x_2{}^2x_4x_5{}^2 + x_2{}^2x_4x_6{}^2 - 2x_2{}^2x_5{}^2x_6 = 0, \\ x_1{}^2x_2{}^2x_6 + 6x_1x_2{}^2x_4x_5 - 6x_1x_2{}^2x_4x_6 - x_1x_4x_5{}^2x_6 + x_1x_4x_5x_6{}^2 \\ \quad +x_2{}^2x_4{}^2x_6 - 2x_2{}^2x_4x_5{}^2 + 2x_2{}^2x_4x_6{}^2 - x_2{}^2x_5{}^2x_6 = 0. \end{cases} \qquad (21)$$

We compute a Gröbner basis for the system of equations (21) by using lex order with $x_1 > x_2 > x_5 > x_6 > x_4$ for $x_1x_2x_4x_5x_6 \neq 0$. Then we obtain the following system of equations:

$$\begin{cases} 6875x_4{}^4 - 11100x_4{}^3 + 7458x_4{}^2 - 2588x_4 + 379 = 0, \\ x_6 - x_4 = 0, \\ -110000x_4{}^3x_5 + 20625x_4{}^3 + 138000x_4{}^2x_5 - 47325x_4{}^2 \\ \quad -58640x_4x_5 + 32247x_4 + 4096x_5{}^2 + 6064x_5 - 6443 = 0, \\ 2048x_2 + 6875x_4{}^3 - 9725x_4{}^2 + 5513x_4 - 1895 = 0, \\ 256x_1 - 6875x_4{}^3 + 8625x_4{}^2 - 3665x_4 + 256x_5 + 379 = 0. \end{cases} \qquad (22)$$

By solving the first equation of (22) for x_4 numerically, we obtain exactly two real solutions which are approximately given by $x_4 \approx 0.58384886$ and $x_4 \approx 0.41501231$. We substitute these values for x_4 into third, fourth and fifth equations of (22), and get four systems of real solutions which are approximately given by

$$\begin{aligned} &1) \quad x_1 \approx 0.147589697, \ x_2 = x_3 \approx 0.30420768, \\ &\qquad x_4 = x_6 \approx 0.58384886, \ x_5 \approx 0.59068145 \\ &2) \quad x_1 \approx 0.59068145, \ x_2 = x_3 \approx 0.304207681, \\ &\qquad x_4 = x_6 \approx 0.58384886, \ x_5 \approx 0.147589697 \\ &3) \quad x_1 \approx 0.15026824, \ x_2 = x_3 \approx 0.38603586, \\ &\qquad x_4 = x_6 \approx 0.41501231, \ x_5 \approx 0.42752296 \\ &4) \quad x_1 \approx 0.42752296, \ x_2 = x_3 \approx 0.386035866, \\ &\qquad x_4 = x_6 \approx 0.41501231, \ x_5 \approx 0.15026824. \end{aligned} \qquad (23)$$

We can see that invariant metrics 1), 2) and 3), 4) are isometric by the action of Weyl group respectively.

Now we consider the case when $x_2 \neq x_3$. We compute a Gröbner basis for the system of equations (20) by using lex order with $x_6 > x_1 > x_3 > x_5 > x_4 > x_2$ for $x_1 x_2 x_3 x_4 x_5 x_6 \neq 0$. Then we obtain the following equation for x_2 in the Gröbner basis:

$$(4x_2 - 1)(8x_2 - 3)\times$$
$$(615007068719846967056793600000000000000x_2^{32}$$
$$-775609679947623330647900160000000000000x_2^{31}$$
$$+459045697424530965142123566858240000000x_2^{30}$$
$$-170858337080972956914577772327731200000000x_2^{29}$$
$$+452037979364735262381167391566462976000000x_2^{28}$$
$$-910329242077850241859282893801521152000000x_2^{27}$$
$$+1457937218786751983356834758166734438400000x_2^{26}$$
$$-1913964445993206156237896958743842652160000x_2^{25}$$
$$+2105011778562298537287259159632835248128000x_2^{24}$$
$$-1971018479819563628158647276861643345100800x_2^{23}$$
$$+1590200280047561979985822623258329753845760x_2^{22}$$
$$-1115363650505052278455943766114909573611520x_2^{21}$$
$$+684603753562702772477174683429198637301760x_2^{20}$$
$$-369460725118424509136799694484827631779840x_2^{19}$$
$$+175873501288501355790284834358020716625920x_2^{18}$$
$$-739928654338846440183994392567939386572800x_2^{17}$$
$$+275371568695288560223025499399619241574400x_2^{16}$$
$$-906440944120552584285407838098058903552000x_2^{15}$$
$$+263624773690615475991324071176446148608000x_2^{14}$$
$$-676037597355704937746908043721540894720000x_2^{13}$$
$$+152388922613987393511973655165355950080000x_2^{12}$$
$$-30067126537437234678956627777756286976000x_2^{11}$$
$$+5163751004230091875288905015728537600000x_2^{10}$$
$$-766406223403654969184652349522575360000x_2^9$$
$$+97408376658495008451656773085429760000x_2^8$$
$$-10478094744093306001891918526873600000x_2^7$$
$$+9395152040402187414912360597504000000x_2^6$$
$$-6881317321099397154873591513600000x_2^5$$
$$+400377368728884170738684080000000x_2^4$$
$$-177685811079067504342510840000000x_2^3$$
$$+56396804056702969555068900000000x_2^2$$
$$-1138083852238036721590500000000x_2$$
$$+1095283909222986099881250000) = 0.$$

(24)

For $(4x_2 - 1)(8x_2 - 3) = 0$, by computing another Gröbner basis for the system of equations (20), we obtain four systems of solutions

$$1) \ x_1 = \frac{1}{8}, \ x_2 = \frac{1}{4}, \ x_3 = \frac{3}{8}, \ x_4 = \frac{1}{2}, \ x_5 = \frac{5}{8}, \ x_6 = \frac{3}{4}$$

$$2) \ x_1 = \frac{5}{8}, \ x_2 = \frac{1}{4}, \ x_3 = \frac{3}{8}, \ x_4 = \frac{1}{2}, \ x_5 = \frac{1}{8}, \ x_6 = \frac{3}{4}$$

$$3) \ x_1 = \frac{1}{8}, \ x_2 = \frac{3}{8}, \ x_3 = \frac{1}{4}, \ x_4 = \frac{3}{4}, \ x_5 = \frac{5}{8}, \ x_6 = \frac{1}{2}$$ \qquad (25)

$$4) \ x_1 = \frac{5}{8}, \ x_2 = \frac{3}{8}, \ x_3 = \frac{1}{4}, \ x_4 = \frac{3}{4}, \ x_5 = \frac{1}{8}, \ x_6 = \frac{1}{2}.$$

These solutions give Kähler Einstein metrics which are isometric by the action of Weyl group to each other.

For degree 32 part of the equation (24), we solve the equation numerically and obtain exactly six real solutions which are approximately given by $x_2 \approx 0.10308692$, $x_2 \approx 0.10468854$, $x_2 \approx 0.34689024$, $x_2 \approx 0.43884703$, $x_2 \approx 0.74009763$, $x_2 \approx 0.83553414$. We also compute a Gröbner basis for the system of equations (20) by using lex order $x_6 > x_1 > x_3 > x_5 > x_4 > x_2$ with $(4x_2 - 1)(8x_2 - 3) \neq 0$ and obtain the system of polynomial equations which have huge coefficients. We substitute these values for x_2 into the system of polynomial equations and then we see that, for $x_2 \approx$ 0.10308692, we have $x_6 \approx -0.22768609$, for $x_2 \approx 0.10468854$, we have $x_6 \approx -0.41413004$, for $x_2 \approx 0.74009763$, we have $x_4 \approx -0.22768609$ and for $x_2 \approx 0.83553414$, we have $x_4 \approx -0.41413004$. For $x_2 \approx 0.34689024$ and $x_2 \approx 0.43884703$, we get four systems of real solutions which are approximately given by

$$1) \ x_1 \approx 0.13872835, x_2 \approx 0.34689024, x_3 \approx 0.43884703,$$
$$x_4 \approx 0.35163915, x_5 \approx 0.43462385, x_6 \approx 0.50132323$$
$$2) \ x_1 \approx 0.43462385, x_2 \approx 0.34689024, x_3 \approx 0.43884703,$$
$$x_4 \approx 0.35163915, x_5 \approx 0.13872835, x_6 \approx 0.50132323$$
$$3) \ x_1 \approx 0.13872835, x_2 \approx 0.43884703, x_3 \approx 0.34689024,$$ \qquad (26)
$$x_4 \approx 0.50132323, x_5 \approx 0.43462385, x_6 \approx 0.35163915$$
$$4) \ x_1 \approx 0.43462385, x_2 \approx 0.43884703, x_3 \approx 0.34689024,$$
$$x_4 \approx 0.50132323, x_5 \approx 0.13872835, x_6 \approx 0.35163915.$$

Note that these are isometric by the action of Weyl group each other.

Case of $x_5 = x_1$. We see that the system of polynomial equations (20) reduces to the following system of polynomial equations:

$$
\begin{cases}
x_1{}^3 - 16x_1{}^2x_2x_3 - x_1x_2{}^2 + 8x_1x_2x_3 - x_1x_3{}^2 \\
\quad -x_2x_3x_4 - x_2x_3x_6 = 0, \\
8x_1x_2{}^2x_3 - x_1x_2{}^2x_6 - 8x_1x_2x_3{}^2 + x_1x_3{}^2x_4 \\
\quad -2x_2{}^3x_3 + 2x_2x_3{}^3 = 0, \\
x_1{}^3x_2x_3x_4 + 4x_1{}^2x_2{}^2x_3{}^2 - 8x_1{}^2x_2{}^2x_3x_4 + x_1{}^2x_2{}^2x_4x_6 \\
\quad +x_1{}^2x_3{}^2x_4{}^2 + x_1x_2{}^3x_3x_4 - x_1x_2x_3{}^3x_4 + x_2{}^2x_3{}^2x_4{}^2 = 0, \\
x_1{}^3x_2x_4 - 4x_1{}^2x_2{}^2x_3 - x_1{}^2x_3x_4{}^2 - x_1x_2{}^3x_4 + 8x_1x_2{}^2x_3x_4 \\
\quad -x_1x_2x_3{}^2x_4 - 2x_2{}^2x_3x_4{}^2 - x_2{}^2x_3x_4x_6 = 0, \\
x_1{}^3x_3x_6 - 4x_1{}^2x_2x_3{}^2 - x_1{}^2x_2x_6{}^2 - x_1x_2{}^2x_3x_6 + 8x_1x_2x_3{}^2x_6 \\
\quad -x_1x_3{}^3x_6 - x_2x_3{}^2x_4x_6 - 2x_2x_3{}^2x_6{}^2 = 0.
\end{cases}
\tag{27}
$$

We compute a Gröbner basis for the system of equations (27) by using lex order with $x_6 > x_1 > x_3 > x_4 > x_2$ for $x_1x_2x_3x_4x_6 \neq 0$. Then we obtain the following equation for x_2 in the Gröbner basis:

$$
\begin{aligned}
&(629145600x_2{}^6 - 1142947840x_2{}^5 + 856522752x_2{}^4 - 339427328x_2{}^3 \\
&+75165440x_2{}^2 - 8838656x_2 + 432181)\times \\
&(15419345916043635437640597110784000000000x_2{}^{30} \\
&-13656400699642646452603688841117696000000x_2{}^{29} \\
&+58756645251594640969735905179231846400000x_2{}^{28} \\
&-163458528754824599182963022452878213120000x_2{}^{27} \\
&+330209849274415583465865408003157524480000x_2{}^{26} \\
&-515799399546076320221574144767872204800000x_2{}^{25} \\
&+647807968029058720451755160832332519178240x_2{}^{24} \\
&-671646144035094400265925101002071075717120x_2{}^{23} \\
&+585751387430835926383471403859641918029824x_2{}^{22} \\
&-435679827066221496967591591028413415030784x_2{}^{21} \\
&+279268844969356415964034853531337011232768x_2{}^{20} \\
&-155496615982662765641584960170545633034240x_2{}^{19} \\
&+75663818256559508584615730115676618096640x_2{}^{18} \\
&-32322437984044118941604502970725065293824x_2{}^{17} \\
&+12162315953026321950643253479064425463808x_2{}^{16} \\
&-4040308640245759618988013385455308374016x_2{}^{15} \\
&+1186518442239198538967334198308304322560x_2{}^{14} \\
&-308159859290703178908267171493548392448x_2{}^{13} \\
&+70745335804350176372577869721537347584x_2{}^{12} \\
&-14334562285684678562476763828084277248x_2{}^{11} \\
&+2556706536220368459513690658455945216x_2{}^{10} \\
&-399779304765182868226126211857776640x_2{}^9
\end{aligned}
\tag{28}
$$

$+5448091189938459552901375354994688 0 x_2{}^8$

$-6416814791215558549153636194713600 x_2{}^7$

$+6454861599992145418449158938624 00 x_2{}^6$

$-545170245196915267131221647360 00 x_2{}^5$

$+377000126453602193402089219200 0 x_2{}^4$

$-205400559063477700662624000000 x_2{}^3$

$+827796085016953221965766000 0 x_2{}^2$

$-2194770958093083808516000 00 x_2 + 2870277947620106110953125)$

$= 0.$

We solve the equation (28) numerically and see that there exist no real solutions. This completes the proof of Theorem 1.1.

8.2. Case of $Sp(4)/(U(1) \times U(1) \times Sp(2))$

For $Sp(4)/(U(1) \times U(1) \times Sp(2))$, the system of equations (19) reduces to the following system of polynomial equations:

$$
\begin{cases}
x_1{}^2 x_2 x_3 x_4 + x_1{}^2 x_2 x_3 x_6 + 2 x_1{}^2 x_4 x_5 x_6 - 20 x_1 x_2 x_3 x_4 x_5 x_6 \\
-2 x_2{}^2 x_4 x_5 x_6 - x_2 x_3 x_4{}^2 x_6 - x_2 x_3 x_4 x_5{}^2 + 10 x_2 x_3 x_4 x_5 x_6 \\
-x_2 x_3 x_4 x_6{}^2 - x_2 x_3 x_5{}^2 x_6 - 2 x_3{}^2 x_4 x_5 x_6 = 0, \\
(x_1 - x_5)\,(x_1 x_2 x_3 x_4 + x_1 x_2 x_3 x_6 + x_1 x_4 x_5 x_6 + x_2{}^2 x_4 x_6 \\
\quad + x_2 x_3 x_4 x_5 - 5 x_2 x_3 x_4 x_6 + x_2 x_3 x_5 x_6 + x_3{}^2 x_4 x_6) = 0, \\
x_1 x_2{}^3 x_3 - 10 x_1 x_2{}^2 x_3 x_5 + x_1 x_2{}^2 x_5 x_6 - x_1 x_2 x_3{}^3 + 10 x_1 x_2 x_3{}^2 x_5 \\
-x_1 x_3{}^2 x_4 x_5 + x_2{}^3 x_3 x_5 - x_2 x_3{}^3 x_5 = 0, \\
2 x_1{}^2 x_2{}^2 x_3{}^2 - x_1{}^2 x_2 x_3 x_4 x_5 - x_1 x_2{}^3 x_3 x_4 - 12 x_1 x_2{}^2 x_3{}^2 x_5 \\
+20 x_1 x_2{}^2 x_3 x_4 x_5 - 2 x_1 x_2{}^2 x_4 x_5 x_6 + x_1 x_2 x_3{}^3 x_4 \\
-x_1 x_2 x_3 x_4 x_5{}^2 - 4 x_1 x_3{}^2 x_4{}^2 x_5 - x_2{}^3 x_3 x_4 x_5 - 2 x_2{}^2 x_3{}^2 x_4{}^2 \\
+2 x_2{}^2 x_3{}^2 x_5{}^2 + x_2 x_3{}^3 x_4 x_5 = 0, \\
x_1{}^2 x_2{}^2 x_3 x_4 + 2 x_1 x_2{}^3 x_4 x_6 - 10 x_1 x_2{}^2 x_3 x_4 x_6 + 6 x_1 x_2{}^2 x_3 x_5 x_6 \\
+2 x_1 x_2 x_3{}^2 x_4 x_6 - 2 x_1 x_2 x_4 x_5{}^2 x_6 + 2 x_1 x_3 x_4{}^2 x_5 x_6 + 2 x_2{}^2 x_3 x_4{}^2 x_6 \\
-x_2{}^2 x_3 x_4 x_5{}^2 + x_2{}^2 x_3 x_4 x_6{}^2 - 2 x_2{}^2 x_3 x_5{}^2 x_6 = 0, \\
-x_1{}^2 x_2 x_3{}^2 x_6 - 2 x_1 x_2{}^2 x_3 x_4 x_6 - 6 x_1 x_2 x_3{}^2 x_4 x_5 + 10 x_1 x_2 x_3{}^2 x_4 x_6 \\
-2 x_1 x_2 x_4 x_5 x_6{}^2 - 2 x_1 x_3{}^3 x_4 x_6 + 2 x_1 x_3 x_4 x_5{}^2 x_6 - x_2 x_3{}^2 x_4{}^2 x_6 \\
+2 x_2 x_3{}^2 x_4 x_5{}^2 - 2 x_2 x_3{}^2 x_4 x_6{}^2 + x_2 x_3{}^2 x_5{}^2 x_6 = 0.
\end{cases}
\tag{29}
$$

Case of $x_5 \neq x_1$. We first consider the case when $x_3 = x_2$. By computing a Gröbner basis for the system of equations (29) by using lex order with $x_1 > x_2 > x_5 > x_6 > x_4$ for $x_1 x_2 x_4 x_5 x_6 \neq 0$, we see that $x_6 - x_4 = 0$. Thus the system of polynomial equations (29) reduce to the following system of polynomial equations:

$$\begin{cases} x_1{}^2x_2{}^2 + x_1{}^2x_4x_5 - 10x_1x_2{}^2x_4x_5 - x_2{}^2x_4{}^2 + 3x_2{}^2x_4x_5 \\ -x_2{}^2x_5{}^2 = 0, \\ 2x_1x_2{}^2 + x_1x_4x_5 - 3x_2{}^2x_4 + 2x_2{}^2x_5, 2x_1{}^2x_2{}^2 - x_1{}^2x_4x_5 \\ -12x_1x_2{}^2x_5 + 20x_1x_2x_4x_5 - 6x_1x_4{}^2x_5 - x_1x_4x_5{}^2 \\ -2x_2{}^2x_4{}^2 + 2x_2{}^2x_5{}^2 = 0, \\ x_1{}^2x_2{}^2 - 6x_1x_2{}^2x_4 + 6x_1x_2{}^2x_5 + 2x_1x_4{}^2x_5 - 2x_1x_4x_5{}^2 \\ +3x_2{}^2x_4{}^2 - 3x_2{}^2x_5{}^2 = 0. \end{cases} \qquad (30)$$

We compute a Gröbner basis for the system of equations (30) by using lex order with $x_1 > x_5 > x_4 > x_2$ for $x_1x_2x_4x_5 \neq 0$. Then we obtain the following system of equations:

$$\begin{cases} 367500x_2{}^4 - 516250x_2{}^3 + 272375x_2{}^2 - 64200x_2 + 5722 = 0, \\ 210000x_2{}^3 - 235000x_2{}^2 + 88500x_2 + 385x_4 - 11444 = 0, \\ 2502500x_2{}^3 - 2378200x_2{}^2 - 60\left(105000x_2{}^3 - 121350x_2{}^2\right. \\ \left. +46175x_2 - 5722\right)x_5 + 706200x_2 + 5775x_5{}^2 - 62942 = 0, \\ 385x_1 - 420000x_2{}^3 + 485400x_2{}^2 - 184700x_2 + 385x_5 \\ +22888 = 0. \end{cases} \qquad (31)$$

By solving the first equation of (31) for x_2 numerically, we obtain exactly two real solutions which are approximately given by $x_2 \approx 0.34370919$ and $x_2 \approx 0.43517033$. We substitute these values for x_2 into second, third and fourth equations of (31), and get four systems of real solutions which are approximately given by

$$\begin{array}{ll} 1) & x_1 \approx 0.68034420, \ x_2 = x_3 \approx 0.34370919, \\ & x_4 = x_6 \approx 0.67731898, \ x_5 \approx 0.11376133 \\ 2) & x_1 \approx 0.11376133, \ x_2 = x_3 \approx 0.34370919, \\ & x_4 = x_6 \approx 0.67731898, \ x_5 \approx 0.68034420 \\ 3) & x_1 \approx 0.34273193, \ x_2 = x_3 \approx 0.43517033, \\ & x_4 = x_6 \approx 0.33276508, \ x_5 \approx 0.12021593 \\ 4) & x_1 \approx 0.12021593, \ x_2 = x_3 \approx 0.43517033, \\ & x_4 = x_6 \approx 0.33276508, \ x_5 \approx 0.34273193. \end{array} \qquad (32)$$

We can see that invariant metrics 1), 2) and 3), 4) are isometric by the action of Weyl group respectively.

Now we consider the case when $x_2 \neq x_3$. We compute a Gröbner basis for the system of equations (29) by using lex order with $x_6 > x_1 > x_3 > x_5 > x_4 > x_2$ for $x_1x_2x_3x_4x_5x_6 \neq 0$. Then we obtain a polynomial equation for x_2 which is of degree 36 and $(5x_2 - 2)(10x_2 - 3) = 0$ in the Gröbner basis. For $(5x_2 - 2)(10x_2 - 3) = 0$, by computing a Gröbner basis for the system of equations (29), we obtain four systems of solutions. These give a Kähler Einstein metric up to isometry:

$$x_1 = \frac{1}{10}, \ x_2 = \frac{3}{10}, \ x_3 = \frac{2}{5}, \ x_4 = \frac{3}{5}, \ x_5 = \frac{7}{10}, \ x_6 = \frac{4}{5}. \tag{33}$$

Now we compute a Gröbner basis for the system of equations (29) under the conditions $x_2 \neq x_3$ and $(5x_2 - 2)(10x_2 - 3) \neq 0$. Then we obtain the system of polynomial equations which have huge coefficients. We solve the polynomial equation for x_2 of degree 36 numerically and substitute these values for x_2 into the system of polynomial equations. Then, up to isometry, we see that

$$\begin{aligned} &x_1 \approx 0.13872835, x_2 \approx 0.34689024, x_3 \approx 0.43884703, \\ &x_4 \approx 0.35163915, x_5 \approx 0.43462385, x_6 \approx 0.50132323. \end{aligned} \tag{34}$$

Case of $x_5 = x_1$. We consider the case of $x_3 = x_2$. Then from $r_2 - r_4 = 0$ we see that $x_6 - x_4 = 0$. Then the system of equations (29) is given by

$$\begin{cases} x_1{}^3 - 10x_1{}^2x_2{}^2 + 3x_1x_2{}^2 - x_2{}^2x_4 = 0, \\ -x_1{}^3x_4 - 4x_1{}^2x_2{}^2 + 10x_1{}^2x_2x_4 - 3x_1{}^2x_4{}^2 - x_2{}^2x_4{}^2 = 0, \\ -2x_1{}^3x_4 + 4x_1{}^2x_2{}^2 + 2x_1{}^2x_4{}^2 - 6x_1x_2{}^2x_4 + 3x_2{}^2x_4{}^2 = 0. \end{cases} \tag{35}$$

We compute a Gröbner basis for the system of equations (35). Then we obtain the following equations:

$$\begin{cases} 4704000000x_2{}^6 - 10136000000x_2{}^5 + 9057640000x_2{}^4 \\ -4298920000x_2{}^3 + 1143653600x_2{}^2 - 161802000x_2 + 9517117 = 0, \\ -750288000000x_2{}^5 + 1091631520000x_2{}^4 - 628293100000x_2{}^3 \\ +181214485800x_2{}^2 - 26661224000x_2 + 12706460x_4 \\ +1627427007 = 0, \\ 12706460x_1 + 750288000000x_2{}^5 - 1091631520000x_2{}^4 \\ +628293100000x_2{}^3 - 180960356600x_2{}^2 + 26534159400x_2 \\ -1627427007 = 0. \end{cases}$$

Solving the equations above numerically, we obtain two systems of positive solutions which are approximately given by

$$\begin{aligned} &1) \ x_1 = x_5 \approx 0.80038772, x_2 = x_3 \approx 0.34660823, \\ &\quad x_4 = x_6 \approx 0.2629492, \\ &2) \ x_1 = x_5 \approx 0.63632178, x_2 = x_3 \approx 0.30995274, \\ &\quad x_4 = x_6 \approx 0.54179159. \end{aligned} \tag{36}$$

We consider the case of $x_3 \neq x_2$. We compute a Gröbner basis for the system of equations (29) by using lex order with $x_6 > x_1 > x_3 > x_4 > x_2$ for $x_1x_2x_3x_4x_6 \neq 0$. Then we obtain a polynomial equation for x_2 of degree

30 in the Gröbner basis and the system of polynomial equations which have huge coefficients. We solve the polynomial equation for x_2 of degree 30 numerically and substitute these values for x_2 into the system of polynomial equations. Then, up to isometry, we see that

$$x_1 = x_5 \approx 0.71312194, \quad x_2 \approx 0.35355789, \quad x_3 \approx 0.30533774,$$
$$x_4 \approx 0.26136378, \quad x_6 \approx 0.53441520 \tag{37}$$

give an Einstein metric on $Sp(4)/(U(2) \times U(1) \times Sp(1))$. This completes the proof of Theorem 1.2.

8.3. Case of $Sp(4)/(U(2) \times U(1) \times Sp(1))$

For $Sp(4)/(U(2) \times U(1) \times Sp(1))$, the system of equations (19) reduces to the following system of polynomial equations:

$$
\begin{cases}
3x_1{}^2x_2x_3x_4 + 2x_1{}^2x_2x_3x_6 + 2x_1{}^2x_4x_5x_6 - 40x_1x_2x_3x_4x_5x_6 \\
-2x_2{}^2x_4x_5x_6 - 2x_2x_3x_4{}^2x_6 - 3x_2x_3x_4x_5{}^2 + 20x_2x_3x_4x_5x_6 \\
-3x_2x_3x_4x_6{}^2 - 2x_2x_3x_5{}^2x_6 - 2x_3{}^2x_4x_5x_6 = 0, \\
(x_1 - x_5)\left(3x_1x_2x_3x_4 + 2x_1x_2x_3x_6 + x_1x_4x_5x_6 + x_2{}^2x_4x_6\right. \\
\left.+3x_2x_3x_4x_5 - 10x_2x_3x_4x_6 + 2x_2x_3x_5x_6 + x_3{}^2x_4x_6\right) = 0, \\
-x_1{}^2x_2x_3x_5 + 3x_1x_2{}^3x_3 - 20x_1x_2{}^2x_3x_5 + 3x_1x_2{}^2x_5x_6 \\
-3x_1x_2x_3{}^3 + 20x_1x_2x_3{}^2x_5 - x_1x_2x_3x_5{}^2 - 2x_1x_3{}^2x_4x_5 \\
+3x_2{}^3x_3x_5 - 3x_2x_3{}^3x_5 = 0, \\
4x_1{}^2x_2{}^2x_3{}^2 - x_1{}^2x_2x_3x_4x_5 - x_1x_2{}^3x_3x_4 - 16x_1x_2{}^2x_3{}^2x_5 \\
+20x_1x_2{}^2x_3x_4x_5 - 3x_1x_2{}^2x_4x_5x_6 + x_1x_2x_3{}^3x_4 - x_1x_2x_3x_4x_5{}^2 \\
-2x_1x_3{}^2x_4{}^2x_5 - x_2{}^3x_3x_4x_5 - 4x_2{}^2x_3{}^2x_4{}^2 + 4x_2{}^2x_3{}^2x_5{}^2 \\
+x_2x_3{}^3x_4x_5 = 0, \\
3x_1{}^2x_2{}^2x_3x_4 - 2x_1{}^2x_2{}^2x_3x_6 + 2x_1x_2{}^3x_4x_6 - 20x_1x_2{}^2x_3x_4x_6 \\
+16x_1x_2{}^2x_3x_5x_6 + 2x_1x_2x_3{}^2x_4x_6 - 2x_1x_2x_4x_5{}^2x_6 \\
+2x_1x_3x_4{}^2x_5x_6 + 6x_2{}^2x_3x_4{}^2x_6 - 3x_2{}^2x_3x_4x_5{}^2 + 3x_2{}^2x_3x_4x_6{}^2 \\
-6x_2{}^2x_3x_5{}^2x_6 = 0, \\
-x_1{}^2x_2x_3{}^2x_4 - 2x_1{}^2x_2x_3{}^2x_6 - 2x_1x_2{}^2x_3x_4x_6 - 16x_1x_2x_3{}^2x_4x_5 \\
+20x_1x_2x_3{}^2x_4x_6 - 2x_1x_2x_4x_5x_6{}^2 - 2x_1x_3{}^3x_4x_6 + 2x_1x_3x_4x_5{}^2x_6 \\
-2x_2x_3{}^2x_4{}^2x_6 + 5x_2x_3{}^2x_4x_5{}^2 - 5x_2x_3{}^2x_4x_6{}^2 \\
+2x_2x_3{}^2x_5{}^2x_6 = 0.
\end{cases}
\tag{38}
$$

Case of $x_5 \neq x_1$. We compute a Gröbner basis for the system of equations (38) by using lex order with $x_6 > x_1 > x_3 > x_4 > x_5 > x_2$ for $x_1x_2x_3x_4x_5x_6 \neq 0$. Then we obtain a polynomial equation for x_2 which is of degree 46 and $(5x_2 - 2)(5x_2 - 1) = 0$ in the Gröbner basis. For $(5x_2 - 2)(5x_2 - 1) = 0$, by computing a Gröbner basis for the system of equations (38), we obtain four systems of solutions

$$1)\ x_1 = \frac{3}{20},\ x_2 = \frac{2}{5},\ x_3 = \frac{1}{4},\ x_4 = \frac{4}{5},\ x_5 = \frac{13}{20},\ x_6 = \frac{1}{2}$$

$$2)\ x_1 = \frac{13}{20},\ x_2 = \frac{2}{5},\ x_3 = \frac{1}{4},\ x_4 = \frac{4}{5},\ x_5 = \frac{3}{20},\ x_6 = \frac{1}{2}$$

$$3)\ x_1 = \frac{3}{20},\ x_2 = \frac{1}{5},\ x_3 = \frac{7}{20},\ x_4 = \frac{2}{5},\ x_5 = \frac{11}{20},\ x_6 = \frac{7}{10}$$

$$4)\ x_1 = \frac{11}{20},\ x_2 = \frac{1}{5},\ x_3 = \frac{7}{20},\ x_4 = \frac{2}{5},\ x_5 = \frac{3}{20},\ x_6 = \frac{7}{10}.$$

(39)

Note that invariant metrics 1), 2) and 3), 4) are isometric by the action of Weyl group respectively and these are Kähler Einstein with respect to two different $Sp(4)$-invariant complex structures on $Sp(4)/(U(2) \times U(1) \times Sp(1))$ (cf. [14]).

For the part of a polynomial equation for x_2 of degree 46 in the Gröbner basis, we solve the polynomial equation for x_2 numerically and substitute these values for x_2 into the system of polynomial equations in the Gröbner basis. Then, up to isometry, we see that

$$\begin{aligned}
1)\ \ &x_1 \approx 0.18068519, x_2 \approx 0.25108317, x_3 \approx 0.31498778,\\
&x_4 \approx 0.46550020, x_5 \approx 0.53572002, x_6 \approx 0.58939892\\
2)\ \ &x_1 \approx 0.50630777, x_2 \approx 0.46123056, x_3 \approx 0.33135468,\\
&x_4 \approx 0.62124963, x_5 \approx 0.15821909, x_6 \approx 0.38146255
\end{aligned}$$

(40)

give Einstein metrics on $Sp(4)/(U(2) \times U(1) \times Sp(1))$.

Case of $x_5 = x_1$. We compute a Gröbner basis for the system of equations (38) by using lex order with $x_6 > x_1 > x_3 > x_4 > x_2$ for $x_1 x_2 x_3 x_4 x_6 \neq 0$. Then we get a polynomial equation for x_2 which is of degree 30 in the Gröbner basis. By solving the polynomial equation numerically, we see that there exist no real solutions. This completes the proof of Theorem 1.3.

Acknowledgments

The third author was supported by Grant-in-Aid for Scientific Research (C) 21540080.

References

1. D. V. Alekseevsky: *Homogeneous Einstein metrics*, Differential Geometry and its Applications (Proccedings of the Conference), 1–21. Univ. of. J. E. Purkyne, Chechoslovakia (1987).

2. D. V. Alekseevsky and A. M. Perelomov: *Invariant Kähler-Einstein metrics on compact homogeneous spaces*, Funct. Anal. Appl. 20 (3) (1986) 171–182.

3. S. Anastassiou and I. Chrysikos: *The Ricci flow approach to homogeneous Einstein metrics on flag manifolds*, to appear in J. Geom. Phys.

4. A. Arvanitoyeorgos: *New invariant Einstein metrics on generalized flag manifolds*, Trans. Am. Math. Soc. 337(2) (1993) 981–995.

5. A. Arvanitoyeorgos and I. Chrysikos: *Invariant Einstein metrics on flag manifolds with four isotropy summands*, Ann. Glob. Anal. Geom. 37 (2) (2010) 185–219.

6. A. Arvanitoyeorgos and I. Chrysikos: *Invariant Einstein metrics on generalized flag manifolds with two isotropy summands*, to appear in J. Aust. Math. Soc.

7. A. Arvanitoyeorgos, I. Chrysikos and Y. Sakane: *Complete description of invariant Einstein metrics on the generalized flag manifold* $SO(2n)/U(p) \times U(n - p)$, Ann. Glob. Anal. Geom. 38 (4) (2010) 413–438.

8. A. Arvanitoyeorgos, I. Chrysikos and Y. Sakane: *Homogeneous Einstein metrics on the generalized flag manifold* $Sp(n)/(U(p) \times U(n - p))$, Differential Geometry and its Applications (DGA2010), doi:10.1016/j.difgeo.2011.04.003.

9. A. Arvanitoyeorgos, I. Chrysikos and Y. Sakane: *Homogeneous Einstein metrics on* G_2/T, preprint, math.arXiv: 1010.3661.

10. A. L. Besse: *Einstein manifolds*, Springer-Verlag, Berlin Heidelberg 1987.

11. I. Chrysikos: *Flag manifolds, symmetric t-triples and Einstein metrics*, preprint, math. arXiv:1010.3992.

12. E.C.F. dos Santos and C.J.C. Negreiros: *Einstein metrics on flag manifolds*, Revista Della, Unión Mathemática Argetina 47(2) (2006) 77–84.

13. M. Kimura: *Homogeneous Einstein metrics on certain Kähler C-spaces*, Adv. Stud. Pure Math. 18 (1) (1990) 303–320.

14. M. Nishiyama: *Classification of invariant complex structures on irreducible compact simply connected coset spaces*, Osaka J. Math. 21 (1984) 39–58.

15. B. O'Neill: *The fundamental equation of a submersion*, Michigan Math. J. 13 (1966) 459–469.

16. J-S. Park and Y. Sakane: *Invariant Einstein metrics on certain homogeneous spaces*, Tokyo J. Math. 20 (1) (1997) 51–61.

17. Y. Sakane: *Homogeneous Einstein metrics on flag manifolds*, Towards 100 years after Sophus Lie (Kazan, 1998), Lobachevskii J. Math. 4 (1999), 71–87.

18. M. Wang and W. Ziller: *Existence and non-existence of homogeneous Einstein metrics*, Invent. Math. 84 (1986) 177–194.

Received January 31, 2011
Revised March 29, 2011

Proceedings of the 2nd International
Colloquium on Differential Geometry
and its Related Fields
Veliko Tarnovo, September 6–10, 2010

ON G_2-INVARIANTS OF CURVES
IN PURELY IMAGINARY OCTONIONS

Misa OHASHI

Department of Mathematics, Meijo University,
Nagoya 468-8502, Japan
E-mail: m0851501@ccalumni.meijo-u.ac.jp

We determine the invariant functions of curves in purely imaginary octonions
Im $\mathfrak{C}$ up to the G_2-congruency and prove a G_2-congruence theorem of such
curves. In particular, we write down G_2-invatiants and G_2-frame field for helices
in $\mathbf{R}^4$.

Keywords: G_2, Frenet-Serre formula, G_2-congruence theorem.

1. Introduction

Let $\gamma : I \to \mathbf{E}^n$ be a regular curve in an n-dimensional Euclidean space
$\mathbf{E}^n$ with the canonical metric. The orientation preserving isometry group
of $\mathbf{E}^n$ is the semi-direct product $\mathbf{R}^n \rtimes SO(n)$. Two regular curves γ, $\tilde{\gamma}$:
$I \to \mathbf{E}^n$ of $\mathbf{E}^n$ are called (orientation preserving) $SO(n)$-congruent to each
other if there exists $(a, g) \in \mathbf{R}^n \rtimes SO(n)$ with $\tilde{\gamma} = g \circ \gamma + a$. It is clear
that if two regular curves are $SO(n)$-congruent to each other then their
series of curvatures coincide, and its converse is also true (under the general
condition).

In the case $n = 8$, besides the geometry of curves on $\mathbf{E}^8$ under the
action of $SO(8)$ we can consider another geometry of them. A Euclidean
8-space $\mathbf{E}^8$ has a special algebraic structures, which is called the octonions
$\mathfrak{C}$ (or the Cayley algebra). The automorphism group of the octonions is the
exceptional simple Lie group G_2 which is a Lie subgroup of $SO(7)$ (see §2).
There exists a faithful representation of G_2 to $End_{\mathbf{R}}(\text{Im } \mathfrak{C})$, where Im $\mathfrak{C}$
denotes the set of purely imaginary octonions. In this paper, we consider
the G_2 geometry (Im $\mathfrak{C}$, Im $\mathfrak{C} \rtimes G_2$), which is a subgeometry of $\mathbf{R}^7 \rtimes SO(7)$.

Two curves γ, $\tilde{\gamma} : I \to \text{Im } \mathfrak{C}$ in Im $\mathfrak{C}$ with the same parameterizations
and orientation are called G_2-congruent to each other if there exists $(a, h) \in$

$\operatorname{Im} \mathfrak{C} \rtimes G_2$ with

$$\tilde{\gamma} = h \circ \gamma + a.$$

The purpose of this paper is the following:

(1) Determine the complete invariants of curves in $\operatorname{Im} \mathfrak{C}$ up to the G_2-congruency;
(2) Prove a G_2-congruence theorem of curves in $\operatorname{Im} \mathfrak{C}$;
(3) Give a method of calculation of complete G_2-invariants of curves in $\operatorname{Im} \mathfrak{C}$ from their standard Frenet-Serre formula;
(4) Give examples of pairs of curves which are $SO(7)$-congruent to each other but not G_2-congruent;
(5) Give an explicit representation of G_2-frame field along helices in 3-dimensional and 4-dimensional Euclidean spaces.

In the present paper, all curves and tensor fields are always assumed to be of class C^∞, unless otherwise specified.

2. Preliminaries

Let $\mathbf{H}$ be the skew field of all quaternions with canonical basis $\{1, i, j, k\}$, satisfying

$$i^2 = j^2 = k^2 = -1, \quad ij = -ji = k, \quad jk = -kj = i, \quad ki = -ik = j.$$

The octonions (or the Cayley) $\mathfrak{C}$ over $\mathbf{R}$ can be considered as a direct sum $\mathbf{H} \oplus \mathbf{H}$ with the following multiplication

$$(a + b\varepsilon)(c + d\varepsilon) = ac - \bar{d}b + (da + b\bar{c})\varepsilon,$$

where $\varepsilon = (0, 1) \in \mathbf{H} \oplus \mathbf{H}$ and $a, b, c, d \in \mathbf{H}$, where for $a \in \mathbf{H}$ we denote by $\bar{a}$ its quaternionic conjugate. For arbitrary $x, y \in \mathfrak{C}$, we have

$$\langle xy, xy \rangle = \langle x, x \rangle \langle y, y \rangle.$$

In [2], this condition is called "normed algebra". The octonions is a non-commutative, non-associative alternative division algebra. The group of automorphisms of the octonions is the exceptional simple Lie group

$$G_2 = \{g \in SO(8) \mid g(uv) = g(u)g(v) \text{ for any } u, v \in \mathfrak{C}\}.$$

The "exterior product" of $\mathfrak{C}$ is defined by

$$u \times v = (1/2)(\bar{v}u - \bar{u}v),$$

where $\bar{v} = 2\langle v, 1 \rangle - v$ is the conjugation of $v \in \mathfrak{C}$. We note that $u \times v \in \operatorname{Im} \mathfrak{C}$, where

$$\operatorname{Im} \mathfrak{C} = \{u \in \mathfrak{C} \mid \langle u, 1 \rangle = 0\}.$$

3. G_2-congruence theorem of curves in $\mathrm{Im}\,\mathfrak{C}$

3.1. G_2-frame field along a curve

Let $\gamma : I \to \mathrm{Im}\,\mathfrak{C}$ be a regular curve with an arc-length parameter $s \in I$. We take the tangent vector field along γ as $e_4(s) = \gamma'(s)$. We set $k_1(s) = \|\gamma''(s)\|$ and assume that this function does not vanish everywhere. Put $e_1(s) = \dfrac{1}{k_1} e_4'(s)$. The symbol " $\prime$ " denotes the derivative of the arc-length parameter s. Next we put

$$e_5(s) = e_1(s)e_4(s),$$

then we find that $\langle e_5, e_i \rangle = 0$ for each $i \in \{1, 4\}$. In this paper, we assume

$$\kappa_2(s) = \sqrt{\|e_1'(s)\|^2 - \langle e_1'(s), e_4(s)\rangle^2 - \langle e_1'(s), e_5(s)\rangle^2} > 0.$$

We can then define $e_2(s)$ as

$$e_2(s) = \frac{1}{\kappa_2(s)}\left(e_1'(s) - \langle e_1'(s), e_4(s)\rangle\, e_4(s) - \langle e_1'(s), e_5(s)\rangle\, e_5(s)\right).$$

Then we see that $\|e_2\| = 1$. Lastly, we set $e_3(s)$, $e_6(s)$ and $e_7(s)$ as

$$e_3(s) = e_1(s)e_2(s), \quad e_6(s) = e_2(s)e_4(s), \quad e_7(s) = e_3(s)e_4(s).$$

The multiplication table of $(e_4\ e_1\ e_2\ e_3\ e_5\ e_6\ e_7)$ coincides with that of $(\varepsilon\ i\ j\ k\ i\varepsilon\ j\varepsilon\ k\varepsilon)$. Therefore there exists an G_2-valued function g satisfying

$$(e_4\ e_1\ e_2\ e_3\ e_5\ e_6\ e_7) = \big(g(\varepsilon)\ g(i)\ g(j)\ g(k)\ g(i\varepsilon)\ g(j\varepsilon)\ g(k\varepsilon)\big).$$

In this way, we obtain G_2-frame field along γ.

Proposition 3.1. *Let $\gamma : I \to \mathrm{Im}\,\mathfrak{C}$ be a curve with k_1, $\kappa_2 > 0$. The associated G_2-frame field $(e_4, e_1, \cdots, e_7)$ along γ satisfies the following differential equation*

$$\frac{d}{ds}(e_4\ e_1\ \cdots e_7) = (e_4\ e_1\ \cdots e_7)M, \tag{1}$$

where

$$M = M(k_1,\ \kappa_2,\ \rho_1,\ \rho_2,\ \rho_3,\ \alpha,\ \beta_1,\ \beta_2)$$

$$= \left(\begin{array}{c|ccc|ccc}
0 & -k_1 & 0 & 0 & 0 & 0 & 0 \\
\hline
k_1 & 0 & -\kappa_2 & 0 & -\rho_1 & 0 & 0 \\
0 & \kappa_2 & 0 & -\alpha & 0 & -\rho_2 & -\beta_1 \\
0 & 0 & \alpha & 0 & 0 & -\beta_2 & -\rho_3 \\
\hline
0 & \rho_1 & 0 & 0 & 0 & -\kappa_2 & 0 \\
0 & 0 & \rho_2 & \beta_2 & \kappa_2 & 0 & -\alpha \\
0 & 0 & \beta_1 & \rho_3 & 0 & \alpha & 0
\end{array}\right) \tag{2}$$

with functions given by $\rho_1 = \langle e'_1, e_5 \rangle$, $\rho_2 = \langle e'_2, e_6 \rangle$, $\rho_3 = \langle e'_3, e_7 \rangle$, $\alpha = \langle e'_2, e_3 \rangle$, $\beta_1 = \langle e'_2, e_7 \rangle$, $\beta_2 = \langle e'_3, e_6 \rangle$. *These functions satisfy the following:*

$$\rho_1 + \rho_2 + \rho_3 = 0, \tag{3}$$
$$\beta_1 - \beta_2 - k_1 = 0. \tag{4}$$

Proof. From the definition of G_2-frame field $(e_4, \cdots, e_7)$, we have

$$e'_4 = k_1 e_1. \tag{5}$$

Since $G_2 \subset SO(7)$, the matrix M in (1) is a $\mathfrak{so}(7)$-valued function which satisfies $M + {}^t M = 0_{7 \times 7}$. Here $0_{7 \times 7} \in M_{7 \times 7}(\mathbf{R})$ is a zero matrix. So we may put

$$M = \left(\begin{array}{c|c|c} 0 & -{}^t\mu & 0_{1 \times 3} \\ \hline \mu & A_1 & -{}^t B \\ \hline 0_{3 \times 1} & B & A_2 \end{array} \right),$$

where $\mu = {}^t\begin{pmatrix} k_1 & 0 & 0 \end{pmatrix}$ is an $M_{3 \times 1}(\mathbf{R})$-valued function and A_1, A_2, and B are $M_{3 \times 3}(\mathbf{R})$-valued functions on I. We shall prove that

$$A_1 = A_2 = \begin{pmatrix} 0 & -\kappa_2 & 0 \\ \kappa_2 & 0 & -\alpha \\ 0 & \alpha & 0 \end{pmatrix}, \quad B = \begin{pmatrix} \rho_1 & 0 & 0 \\ 0 & \rho_2 & \beta_2 \\ 0 & \beta_1 & \rho_3 \end{pmatrix}. \tag{6}$$

First we shall show that

$$e'_1 = -k_1 e_4 + \kappa_2 e_2 + \rho_1 e_5, \tag{7}$$
$$e'_2 = -\kappa_2 e_1 + \alpha e_3 + \rho_2 e_6 + \beta_1 e_7, \tag{8}$$
$$e'_3 = -\alpha e_2 + \beta_2 e_6 + \rho_3 e_7. \tag{9}$$

We note that $\bar{e}_i = -e_i$ for all $i \in \{1, \ldots, 7\}$ because $T_{\gamma(s)} \operatorname{Im} \mathfrak{C} \cong \operatorname{Im} \mathfrak{C}$. Since $\langle e_2, e_5 \rangle = 0$ by the definition of e_2, with (5) and (7), we have

$$\begin{aligned}
\langle e'_2, e_5 \rangle &= -\langle e_2, e'_5 \rangle = -\langle e_2, (e_1 e_4)' \rangle \\
&= -\langle e_2, e'_1 e_4 \rangle - \langle e_2, e_1 e'_4 \rangle \\
&= -\langle e_2 \bar{e}_4, e'_1 \rangle - \langle \bar{e}_1 e_2, e'_4 \rangle \\
&= \langle e_6, e'_1 \rangle + \langle e_3, e'_4 \rangle \\
&= \langle e_6, (-k_1 e_4 + \kappa_2 e_2 + \rho_1 e_5) \rangle + \langle e_3, k_1 e_1 \rangle = 0.
\end{aligned}$$

Therefore we get (8). In the same way we have (9). By (7), (8) and (9), we obtain (6). For arbitrary $i, j \in \{1, 2, 3\}$, we have

$$\langle e_i, e_j \rangle = \langle e_i e_4, e_j e_4 \rangle = \langle e_{i+4}, e_{j+4} \rangle.$$

Hence $A_1 = A_2$. Next we prove that (3), (4).

$$\rho_3 = \langle e_3', e_7 \rangle = \langle (e_1 e_2)', e_7 \rangle = \langle e_1' e_2, e_7 \rangle + \langle e_1 e_2', e_7 \rangle$$
$$= \langle e_1', e_7 \bar{e}_2 \rangle + \langle e_2', \bar{e}_1 e_7 \rangle = - \langle e_1', e_5 \rangle - \langle e_2', e_6 \rangle = -\rho_1 - \rho_2.$$

So we get (3). In the same way we get

$$\beta_2 = \langle e_3', e_6 \rangle = - \langle e_1', e_4 \rangle + \langle e_2', e_7 \rangle = -k_1 + \beta_1.$$

Thus we have (4). □

Remark 3.1. For a regular curve $\gamma : I \to \operatorname{Im} \mathfrak{C}$, we define a frame field $(v_1 \cdots v_7)$ inductively by $v_1 = \gamma'$, $v_2 = (1/k_1) v_1'$ with $k_1 = \|v_1'\|$ (> 0) and

$$v_i = (1/k_{i-1}) \left\{ v_{i-1}' - \langle v_{i-1}', v_{i-2} \rangle v_{i-2} \right\},$$

by assuming

$$k_{i-1} = \sqrt{\|v_{i-1}'\|^2 - \langle v_{i-1}', v_{i-2} \rangle^2} > 0, \quad (i = 3, \cdots, 6).$$

Then the standard Frenet-Serre formula of $\mathbf{R}^7$ is given by

$$\frac{d}{ds}(v_1 \cdots v_7) = (v_1 \cdots v_7) \begin{pmatrix} 0 & -k_1 & 0 & \cdots & 0 \\ k_1 & 0 & -k_2 & & \vdots \\ 0 & k_2 & 0 & \ddots & 0 \\ \vdots & & \ddots & \ddots & -k_6 \\ 0 & \cdots & 0 & k_6 & 0 \end{pmatrix}.$$

3.2. G_2-invariants

In this section, we give G_2-invariants of curves in $\operatorname{Im} \mathfrak{C}$.

Definition 3.1. Let γ_1, $\gamma_2 : I \to \operatorname{Im} \mathfrak{C}$ be two oriented curves which are parameterized by their arc-lengths. They are said to be G_2-congruent to each other if there exists an element $(g, a) \in G_2 \times \operatorname{Im} \mathfrak{C}$ satisfying

$$g(\gamma_1(s)) + a = \gamma_2(s),$$

for all $s \in I$.

We will show the following.

Proposition 3.2. The 6 functions k_1, κ_2, $\cdots$, β_1 associated with γ in $\operatorname{Im} \mathfrak{C}$ given in Proposition 3.1 are invariants under the action of G_2.

Proof. We assume that curves γ_1, $\gamma_2 : I \to \text{Im}\,\mathfrak{C}$ are G_2-congruent to each other, that is, there exists an element $(g, a) \in G_2 \times \text{Im}\,\mathfrak{C}$ satisfying

$$\gamma_2(s) = g(\gamma_1(s)) + a = (0 \; ; \; \varepsilon\, i \, \cdots \, k\varepsilon) \left(\begin{array}{c|c} 1 & 0_{1\times 7} \\ \hline a & A \end{array} \right) \left(\begin{array}{c} 1 \\ \hline \hat{\gamma}_1(s) \end{array} \right)$$

for all $s \in I$. Here $A \in G_2(\subset M_{7\times 7}(\mathbf{R}))$, and $\gamma_1(s) = (\varepsilon\, i \, \cdots \, k\varepsilon)\, \hat{\gamma}_1(s)$, where $\hat{\gamma}_1$ is an $M_{7\times 1}(\mathbf{R})$-valued function. Let $\{e_i\}$ (resp. $\{\hat{e}_i\}$) be the G_2 frame field along γ_1 (resp. γ_2). Then we can see that

$$(\hat{e}_4\, \hat{e}_1 \, \cdots \, \hat{e}_7) = (e_4\, e_1 \, \cdots \, e_7)\, A = (g(e_4)\, g(e_1) \, \cdots \, g(e_7)).$$

Since $g \in G_2(\subset SO(7))$, we have

$$\langle \hat{e}_i', \hat{e}_j \rangle = \langle g(e_i'), g(e_j) \rangle = \langle e_i', e_j \rangle,$$

for any i, $j \in \{1, \ldots, 7\}$. We get the desired result. $\square$

3.3. G_2-congruence theorem

In this section the above 6 functions in Proposition 3.2 are complete G_2-invariants, that is, the converse of the Proposition 3.2 is also true. Namely we show the following:

Theorem 3.1. *Let γ_1, $\gamma_2 : I \to \text{Im}\,\mathfrak{C}$ be two curves with the same arc-length parameter $s \in I$. If the G_2-invariants $(k_1, \kappa_2, \rho_1, \rho_2, \alpha, \beta_1)$ associated with γ_1 and those $(\tilde{k}_1, \tilde{\kappa}_2, \tilde{\rho}_1, \tilde{\rho}_2, \tilde{\alpha}, \tilde{\beta}_1)$ associated with γ_2 satisfy*

$$k_1 = \tilde{k}_1, \; \kappa_2 = \tilde{\kappa}_2, \; \rho_1 = \tilde{\rho}_1, \; \rho_2 = \tilde{\rho}_2, \; \alpha = \tilde{\alpha}, \; \beta_1 = \tilde{\beta}_1,$$

then these curves are G_2-congruent to each other.

Proof. Let e_i (resp. $\hat{e}_i$) be the G_2-frame field along γ_1 (resp. γ_2). We set the $G_2 \ltimes \text{Im}\,\mathfrak{C}$-valued functions $\tilde{\gamma}_1$, $\tilde{\gamma}_2$ as

$$\tilde{\gamma}_1(s) = (\gamma_1 \; ; \; e_4\, e_1 \, \cdots \, e_7) = (0 \; ; \; \varepsilon\, i \, \cdots \, k\varepsilon) \left(\begin{array}{c|c} 1 & 0_{1\times 7} \\ \hline \hat{\gamma}_1 & A(s) \end{array} \right),$$

$$\tilde{\gamma}_2(s) = (\gamma_2 \; ; \; \tilde{e}_4\, \tilde{e}_1 \, \cdots \, \tilde{e}_7) = (0 \; ; \; \varepsilon\, i \, \cdots \, k\varepsilon) \left(\begin{array}{c|c} 1 & 0_{1\times 7} \\ \hline \hat{\gamma}_2 & B(s) \end{array} \right).$$

Then there exists $G_2 \ltimes \text{Im}\,\mathfrak{C}$-valued function $\tilde{g}$ satisfying

$$\tilde{\gamma}_2(s) = \tilde{g}(s) \cdot \tilde{\gamma}_1(s)$$

for all $s \in I$. We show that

$$\tilde{g}' = (\tilde{\gamma}_2 \cdot (\tilde{\gamma}_1)^{-1})' \equiv 0.$$

Since the complete G_2-invariants of γ_1 and γ_2 concide, we have

$$
\begin{aligned}
\tilde{g}' &= (\tilde{\gamma}_2)' \cdot (\tilde{\gamma}_1)^{-1} + \tilde{\gamma}_2 \cdot ((\tilde{\gamma}_1)^{-1})' \\
&= (\tilde{\gamma}_2)' \cdot (\tilde{\gamma}_1)^{-1} - \tilde{\gamma}_2 \cdot (\tilde{\gamma}_1)^{-1} \tilde{\gamma}_1'(\tilde{\gamma}_1)^{-1} \\
&= \tilde{\gamma}_2 \big(M(\tilde{k}_1, \ \cdots, \ \tilde{\beta}_2) - M(k_1, \ \cdots, \ \beta_2) \big) (\tilde{\gamma}_1)^{-1} = 0.
\end{aligned}
$$

We get the desired result. □

We call the 6 functions $(k_1, \ \kappa_2, \ \rho_1, \ \rho_2, \ \alpha, \ \beta_1)$ associated with γ : $I \to \mathrm{Im}\,\mathfrak{C}$ the complete G_2-invariants. There are other families of complete G_2-invariants, but when we say "the complete G_2-invariants", it means the family of these 6 functions. The relationship between G_2-frame field $(e_4, \ \cdots, \ e_7)$ and a Frenet frame field $(v_1, \ \cdots, \ v_7)$ is given by

$$
e_4 = v_1, \ e_1 = v_2, \ e_5 = v_2 \times v_1, \ e_2 = \frac{1}{\kappa_2}(k_2 v_3 - \rho_1 v_2 \times v_1),
$$

$$
e_3 = \frac{1}{\kappa_2}(k_2 v_2 \times v_3 + \rho_1 v_1), \ e_6 = \frac{1}{\kappa_2}(k_2 v_3 \times v_1 + \rho_1 v_2), \tag{10}
$$

$$
e_7 = \frac{k_2}{\kappa_2}(v_2 \times v_3) \times v_1.
$$

We note that the G_2-frame field along γ depends only on 1st, 2nd and 3rd derivatives of γ. Therefore we obtain

Proposition 3.3. *The following sets of 6 functions are G_2-invariants of a curve γ in $\mathrm{Im}\,\mathfrak{C}$.*

$$
(1) \begin{cases} k_1, \ \kappa_2, \\ \langle v_3, v_2 \times v_1 \rangle, \ \langle v_4, v_3 \times v_1 \rangle, \ \langle v_4, v_3 \times v_2 \rangle, \ \langle v_4, (v_3 \times v_2) \times v_1 \rangle, \end{cases}
$$

$$
(2) \begin{cases} k_1, \ \kappa_2, \\ \left\langle \gamma''', \gamma'' \times \gamma' \right\rangle, \ \left\langle \gamma^{(4)}, \gamma''' \times \gamma' \right\rangle, \\ \left\langle \gamma^{(4)}, \gamma''' \times \gamma'' \right\rangle, \ \left\langle \gamma^{(4)}, (\gamma''' \times \gamma'') \times \gamma' \right\rangle. \end{cases}
$$

The complete G_2-invariants $(k_1, \kappa_2, \cdots, \beta_1)$ can be calculated by one of these sets of functions.

Proof. First, we assume the 6-functions $k_1, \ \kappa_2, \ \rho_1, \ \rho_2, \ \alpha, \ \beta_1$ are given. Then we show the functions in (1) are written by them. Since $\|v_3\| = 1$, we have

$$
\kappa_2 = \sqrt{k_2{}^2 - \rho_1{}^2}. \tag{11}
$$

From Frenet-Serre formula given in Remark 3.1 and from (10), we find

$$\begin{aligned}
\rho_1 &= \langle e_1', e_5 \rangle = \langle v_2', v_2 \times v_1 \rangle \\
&= \langle -k_1 v_1 + k_2 v_3, \ v_2 \times v_1 \rangle = k_2 \langle v_3, \ v_2 \times v_1 \rangle.
\end{aligned} \tag{12}$$

By (10), we see

$$e_2' = -\frac{k_2{}^2}{\kappa_2} v_2 + \left(\frac{k_2}{\kappa_2}\right)' v_3 + \frac{k_2 k_3}{\kappa_2} v_4 - \left(\frac{\rho_1}{\kappa_2}\right)' (v_2 \times v_1) - \frac{k_2 \rho_1}{\kappa_2} (v_3 \times v_1).$$

Thus we have

$$\begin{aligned}
\rho_2 &= \langle e_2', e_6 \rangle \\
&= -\frac{k_2{}^3}{\kappa_2{}^2} \langle v_2, v_3 \times v_1 \rangle + \frac{k_2{}^2 k_3}{\kappa_2{}^2} \langle v_4, v_3 \times v_1 \rangle - \frac{k_2{}^2 \rho_1}{\kappa_2{}^2} \langle v_3 \times v_1, v_3 \times v_1 \rangle \\
&\quad + \frac{k_2{}^2 \rho_1}{\kappa_2{}^2} \langle v_2, v_2 \rangle - \frac{k_2 \rho_1{}^2}{\kappa_2{}^2} \langle v_3 \times v_1, v_2 \rangle \\
&= \frac{k_2}{\kappa_2{}^2} (k_2{}^2 + \rho_1{}^2) \langle v_3, v_2 \times v_1 \rangle - 2\frac{k_2{}^2 \rho_1}{\kappa_2{}^2} + \frac{k_2{}^2 k_3}{\kappa_2{}^2} \langle v_4, v_3 \times v_1 \rangle \\
&= -\frac{\rho_1}{\kappa_2{}^2} (k_2{}^2 - \rho_1{}^2) + \frac{k_2{}^2 k_3}{\kappa_2{}^2} \langle v_4, v_3 \times v_1 \rangle \\
&= -\rho_1 + \frac{k_2{}^2 k_3}{\kappa_2{}^2} \langle v_4, v_3 \times v_1 \rangle.
\end{aligned} \tag{13}$$

In the same way, we can show that

$$\alpha = -\frac{k_2{}^2 k_3}{\kappa_2{}^2} \langle v_4, v_3 \times v_2 \rangle, \tag{14}$$

$$\beta_1 = -\frac{k_2{}^2 k_3}{\kappa_2{}^2} \langle v_4, (v_3 \times v_2) \times v_1 \rangle. \tag{15}$$

In order to show our assertion, we need to prove that third curvature k_3 is obtained by k_2, κ_2, α, β_1, ρ_3 $(= -(\rho_1 + \rho_2))$. Since we have $v_3 = (\kappa_2/k_2) e_2 + (\rho_1/k_2) e_5$, we get

$$\begin{aligned}
-k_2 v_2 + k_3 v_4 &= \left(\frac{\kappa_2}{k_2}\right)' e_2 + \frac{\kappa_2}{k_2} e_2' + \left(\frac{\rho_1}{k_2}\right)' e_5 + \frac{\rho_1}{k_2} e_5' \\
&= \left(\frac{\kappa_2}{k_2}\right)' e_2 + \frac{\kappa_2}{k_2} (-\kappa_2 e_1 + \alpha e_3 + \rho_2 e_6 + \beta_1 e_7) \\
&\quad + \left(\frac{\rho_1}{k_2}\right)' e_5 + \frac{\rho_1}{k_2} (-\rho_1 e_1 + \kappa_2 e_6).
\end{aligned}$$

By (10), (11),

$$k_3 v_4 = \left(\frac{\kappa_2}{k_2}\right)' e_2 + \left(\frac{\rho_1}{k_2}\right)' e_5 + \frac{\kappa_2}{k_2}(\alpha e_3 - \rho_3 e_6 + \beta_1 e_7).$$

So we obtain

$$k_3 = \sqrt{\{\left(\frac{\kappa_2}{k_2}\right)'\}^2 + \{\left(\frac{\rho_1}{k_2}\right)'\}^2 + \left(\frac{\kappa_2}{k_2}\right)^2 (\alpha^2 + \rho_3{}^2 + \beta_1{}^2)} \qquad (16)$$

$$\left(= \sqrt{\{\left(\frac{\kappa_2}{k_2}\right)'\}^2 + \{\left(\frac{\rho_1}{k_2}\right)'\}^2 + \left(\frac{\kappa_2}{k_2}\right)^2 \|e_7'\|^2}\right).$$

From (11), $\cdots$, (16), we get the desired result. We shall show that functions in (1) are replaced by functions in (2), and vice versa. Since

$$\gamma' = v_1,$$
$$\gamma'' = k_1 v_2,$$
$$\gamma''' = k_1' v_2 + k_1 v_2' = -k_1{}^2 v_1 + k_1' v_2 + k_1 k_2 v_3,$$
$$\gamma^{(4)} = -2k_1 k_1' v_1 - k_1{}^2 v_1' + k_1'' v_2 + k_1' v_2' + (k_1 k_2)' v_3 + k_1 k_2 v_3'$$
$$= -3k_1 k_1' v_1 + (k_1'' - k_1{}^3 - k_1 k_2{}^2) v_2$$
$$+ (2k_1' k_2 + k_1 k_2') v_3 + k_1 k_2 k_3 v_4.$$

We get the desired result. $\qquad\square$

We define the associative angle σ of a curve γ as

$$\cos \sigma = \langle v_3, v_2 \times v_1 \rangle.$$

If $\sigma(s) = 0$ (mod. π), then the distribution D which is defined by $D(s) = \mathrm{span}_{\mathbf{R}}\{v_1(s), v_2(s), v_3(s)\}$ along γ is an associative 3-plane.

4. Curves in 3-dimensional Euclidean space $V^3 \subset \mathrm{Im}\,\mathfrak{C}$

Let V^3 be a 3-dimensional Euclidean space of $\mathrm{Im}\,\mathfrak{C}$. Then there exist $\theta \in S^1$ and $g \in G_2$ such that

$$V^3 = \mathrm{span}_{\mathbf{R}}\{g(i), g(j), g(\cos \theta\, k + \sin \theta\, \varepsilon)\}.$$

Let $\gamma : I \to V^3 (\subset \mathrm{Im}\,\mathfrak{C})$ be an arbitrary curve in V^3. Corresponding to the above $g \in G_2$ we see there exists $\gamma_\theta : I \to V_\theta^3$ with $\gamma = g \circ \gamma_\theta$, where $\theta \in S^1$ and $V_\theta^3 = \mathrm{span}_{\mathbf{R}}\{i, j, \cos \theta\, k + \sin \theta\, \varepsilon\}$.

Proposition 4.1. *For a curve in $V^3(\subset \mathrm{Im}\,\mathfrak{C})$, its associative angle is constant.*

Proof. Since γ_θ is included in a 3-dimensional vector space, by (16), we have

$$(\sigma_\theta')^2 + \sin^2\theta \, \|e_7'\|^2 \equiv 0.$$

Therefore the associative angle σ_θ is constant. □

Conversely, let $\gamma : I \to \mathrm{Im}\,\mathfrak{C}$ be a curve in $\mathrm{Im}\,\mathfrak{C}$ with constant associative angle σ and with constant vector e_7. Then there exists a 3-dimensional Euclidean space $V^3 \subset \mathrm{Im}\,\mathfrak{C}$ containing the image $\gamma(I)$. In fact, since the third curvature k_3 of γ is given by

$$k_3 = \sqrt{(\sigma')^2 + \sin^2\sigma \, \|e_7'\|^2},$$

if $\sigma \neq 0$ and e_7 is constant vector then we see that $k_3 = 0$. If $\sigma = 0$ then $\kappa_2 = 0$. Therefore e_i ($i = 2, 3, 6, 7$) cannot be defined. In this case, also $k_3 = 0$.

5. Curves in 4-dimensional Euclidean space $V^4 \subset \mathrm{Im}\,\mathfrak{C}$

Let V^4 be a 4-dimensional Euclidean space of $\mathrm{Im}\,\mathfrak{C}$. There exist $\theta \in S^1$ and $g \in G_2$ such that

$$V^4 = \mathrm{span}_{\mathbf{R}}\{g(-\sin\theta \, k + \cos\theta \, \varepsilon), \, g(i\varepsilon), \, g(j\varepsilon), \, g(k\varepsilon)\}.$$

Let $\gamma : I \to V^4$ be a curve in V^4. Then there exist $g \in G_2$ and $\gamma_\theta : I \to V_\theta^4$ with $\gamma = g \circ \gamma_\theta$, where $\theta \in S^1$ and $V_\theta^4 = \mathrm{span}_{\mathbf{R}}\{-\sin\theta \, k + \cos\theta \, \varepsilon, \, i\varepsilon, \, j\varepsilon, \, k\varepsilon\}$.

Proposition 5.1. *The G_2-invariants of curve γ in 4-dimensional vector space are the following five functions:*

$$k_1(> 0), \quad k_2(> 0), \quad k_3(\neq 0),$$
$$\cos\sigma = \sin\theta \, \det A, \quad \langle v_4, (v_3 \times v_2) \times v_1 \rangle = \cos\theta,$$

where $v_i = (-\sin\theta \, k + \cos\theta \, \varepsilon)a_{1i} + i\varepsilon \, a_{2i} + j\varepsilon \, a_{3i} + k\varepsilon \, a_{4i}$ *and*

$$A = \begin{pmatrix} a_{11} & a_{12} & a_{13} \\ a_{21} & a_{22} & a_{23} \\ a_{31} & a_{32} & a_{33} \end{pmatrix}.$$

Proof. By (16), the function k_3 is considered as a G_2-invariant. Then we will show that $\langle v_4, v_3 \times v_1 \rangle$, $\langle v_4, v_2 \times v_3 \rangle$ are rewritten by k_1, k_2, k_3, and

$\cos \sigma$. As

$$
\begin{aligned}
(\cos \sigma)' &= \langle v_3, v_2 \times v_1 \rangle' \\
&= \langle v_3', v_2 \times v_1 \rangle + \langle v_3, v_2' \times v_1 \rangle + \langle v_3, v_2 \times v_1' \rangle \\
&= \langle (-k_2 v_2 + k_3 v_4), v_2 \times v_1 \rangle \\
&\quad + \langle v_3, (-k_1 v_1 + k_2 v_3) \times v_1 \rangle + \langle v_3, v_2 \times (k_1 v_2) \rangle \\
&= k_3 \langle v_4, v_2 \times v_1 \rangle,
\end{aligned}
$$

we find

$$
\langle v_4, v_2 \times v_1 \rangle = \frac{(\cos \sigma)'}{k_3}.
$$

From this we have

$$
\begin{aligned}
\left(\frac{(\cos \sigma)'}{k_3} \right)' &= \langle v_4, v_2 \times v_1 \rangle' \\
&= \langle v_4', v_2 \times v_1 \rangle + \langle v_4, v_2' \times v_1 \rangle + \langle v_4, v_2 \times v_1' \rangle \\
&= -k_3 \cos \sigma + k_2 \langle v_4, v_3 \times v_1 \rangle.
\end{aligned}
$$

In the same way we can obtain

$$
\begin{aligned}
\langle v_4, v_3 \times v_1 \rangle' &= -k_2 \langle v_4, v_2 \times v_1 \rangle + k_1 \langle v_4, v_3 \times v_2 \rangle \\
&= -\frac{k_2}{k_3} (\cos \sigma)' + k_1 \langle v_4, v_3 \times v_2 \rangle.
\end{aligned}
$$

Next we prove that the associative angle σ satisfies

$$
\cos \sigma = \sin \theta \ \det A.
$$

We give the multiplication table of exterior product of V^4:

$\times$	η_1	$i\varepsilon$	$j\varepsilon$	$k\varepsilon$
η_1	0	$-\eta_2$	$-\eta_3$	$-\eta_4$
$i\varepsilon$	η_2	0	$-k$	j
$j\varepsilon$	η_3	k	0	$-i$
$k\varepsilon$	η_4	$-j$	i	0

where

$$
\eta_1 = -\sin \theta \ k + \cos \theta \ \varepsilon, \qquad \eta_2 = -\sin \theta \ j\varepsilon - \cos \theta \ i,
$$
$$
\eta_3 = \sin \theta \ i\varepsilon - \cos \theta \ j, \qquad \eta_4 = -\sin \theta \ \varepsilon - \cos \theta \ k.
$$

From the above table, we have

$$
\begin{aligned}
v_2 \times v_1 =& i\{(a_{11}a_{22} - a_{21}a_{12})\cos\theta + (a_{41}a_{32} - a_{31}a_{42})\} \\
&+ j\{(a_{11}a_{32} - a_{31}a_{12})\cos\theta + (a_{21}a_{42} - a_{41}a_{22})\} \\
&+ k\{(a_{11}a_{42} - a_{41}a_{12})\cos\theta + (a_{31}a_{22} - a_{21}a_{32})\} \\
&+ \varepsilon(a_{11}a_{42} - a_{41}a_{12})\sin\theta + i\varepsilon(a_{31}a_{12} - a_{11}a_{32})\sin\theta \\
&+ j\varepsilon(a_{11}a_{22} - a_{21}a_{12})\sin\theta.
\end{aligned} \tag{17}
$$

Therefore we get

$$
\begin{aligned}
\cos\sigma =& \langle v_3, v_2 \times v_1 \rangle \\
=& - a_{13}\sin\theta\{(a_{11}a_{42} - a_{41}a_{12})\cos\theta + (a_{31}a_{22} - a_{21}a_{32})\} \\
&+ a_{13}\cos\theta(a_{11}a_{42} - a_{41}a_{12})\sin\theta \\
&+ a_{23}(a_{31}a_{12} - a_{11}a_{32})\sin\theta + a_{33}(a_{11}a_{22} - a_{21}a_{12})\sin\theta \\
=& \sin\theta \begin{vmatrix} a_{11} & a_{12} & a_{13} \\ a_{21} & a_{22} & a_{23} \\ a_{31} & a_{32} & a_{33} \end{vmatrix}.
\end{aligned}
$$

Next we shall prove $\langle v_4, (v_2 \times v_3) \times v_1 \rangle = -\cos\theta$. To do this, we prepare the following algebraic identity;

$$
\begin{aligned}
& \begin{vmatrix} \langle v_1, v_2 \rangle & \langle v_1, v_3 \rangle \\ \langle v_4, v_2 \rangle & \langle v_4, v_3 \rangle \end{vmatrix} \\
&= \left(\sum_{i=1}^{4} a_{i1}a_{i2}\right)\left(\sum_{i=1}^{4} a_{i4}a_{i3}\right) - \left(\sum_{i=1}^{4} a_{i1}a_{i3}\right)\left(\sum_{i=1}^{4} a_{i4}a_{i2}\right) \\
&= \begin{vmatrix} a_{11} & a_{14} \\ a_{21} & a_{24} \end{vmatrix}\begin{vmatrix} a_{12} & a_{13} \\ a_{22} & a_{23} \end{vmatrix} + \begin{vmatrix} a_{31} & a_{34} \\ a_{41} & a_{44} \end{vmatrix}\begin{vmatrix} a_{32} & a_{33} \\ a_{42} & a_{43} \end{vmatrix} + \begin{vmatrix} a_{11} & a_{14} \\ a_{31} & a_{34} \end{vmatrix}\begin{vmatrix} a_{12} & a_{13} \\ a_{32} & a_{33} \end{vmatrix} \\
&\quad + \begin{vmatrix} a_{21} & a_{24} \\ a_{41} & a_{44} \end{vmatrix}\begin{vmatrix} a_{22} & a_{23} \\ a_{42} & a_{43} \end{vmatrix} + \begin{vmatrix} a_{11} & a_{14} \\ a_{41} & a_{44} \end{vmatrix}\begin{vmatrix} a_{12} & a_{13} \\ a_{42} & a_{43} \end{vmatrix} + \begin{vmatrix} a_{21} & a_{24} \\ a_{31} & a_{34} \end{vmatrix}\begin{vmatrix} a_{22} & a_{32} \\ a_{23} & a_{33} \end{vmatrix} \\
&= \langle v_1 \wedge v_4, v_2 \wedge v_3 \rangle.
\end{aligned}
$$

We note that if $\{v_1, v_2, v_3, v_4\}$ is an orthonormal basis of $\mathbf{R}^4$ then $\langle v_1 \wedge v_4, v_2 \wedge v_3 \rangle = 0$ holds. Since $(v_1, \cdots, v_4)$ is an orthonormal frame field, we see

$$
\langle v_4, (v_2 \times v_3) \times v_1 \rangle = \langle v_4 \times \bar{v}_1, v_2 \times v_3 \rangle = \langle v_1 \times v_4, v_2 \times v_3 \rangle.
$$

In the same way as the calculation in (17), we obtain the following:

$$
\begin{aligned}
v_1 \times v_4 ={}& i\{(a_{14}a_{21} - a_{24}a_{11})\cos\theta + (a_{44}a_{31} - a_{34}a_{41})\} \\
& + j\{(a_{14}a_{31} - a_{34}a_{11})\cos\theta + (a_{24}a_{41} - a_{44}a_{21})\} \\
& + k\{(a_{14}a_{41} - a_{44}a_{11})\cos\theta + (a_{34}a_{21} - a_{24}a_{31})\} \\
& + \varepsilon(a_{14}a_{41} - a_{44}a_{11})\sin\theta + i\varepsilon(a_{34}a_{11} - a_{14}a_{31})\sin\theta \\
& + j\varepsilon(a_{14}a_{21} - a_{24}a_{11})\sin\theta, \\
v_2 \times v_3 ={}& i\{(a_{13}a_{22} - a_{23}a_{12})\cos\theta + (a_{43}a_{32} - a_{33}a_{42})\} \\
& + j\{(a_{13}a_{32} - a_{33}a_{12})\cos\theta + (a_{23}a_{42} - a_{43}a_{22})\} \\
& + k\{(a_{13}a_{42} - a_{43}a_{12})\cos\theta + (a_{33}a_{22} - a_{23}a_{32})\} \\
& + \varepsilon(a_{13}a_{42} - a_{43}a_{12})\sin\theta + i\varepsilon(a_{33}a_{12} - a_{13}a_{32})\sin\theta \\
& + j\varepsilon(a_{13}a_{22} - a_{23}a_{12})\sin\theta.
\end{aligned}
$$

Hence we get

$$
\langle v_1 \times v_4, v_2 \times v_3 \rangle
$$

$$
= -\cos\theta
\begin{vmatrix}
a_{11} & a_{12} & a_{13} & a_{14} \\
a_{21} & a_{22} & a_{23} & a_{24} \\
a_{31} & a_{32} & a_{33} & a_{34} \\
a_{41} & a_{42} & a_{43} & a_{44}
\end{vmatrix}
$$

$$
+ \begin{vmatrix} a_{11} & a_{14} \\ a_{21} & a_{24} \end{vmatrix}
\begin{vmatrix} a_{12} & a_{13} \\ a_{22} & a_{23} \end{vmatrix}
+ \begin{vmatrix} a_{31} & a_{34} \\ a_{41} & a_{44} \end{vmatrix}
\begin{vmatrix} a_{32} & a_{33} \\ a_{42} & a_{43} \end{vmatrix}
+ \begin{vmatrix} a_{11} & a_{14} \\ a_{31} & a_{34} \end{vmatrix}
\begin{vmatrix} a_{12} & a_{13} \\ a_{32} & a_{33} \end{vmatrix}
$$

$$
+ \begin{vmatrix} a_{21} & a_{24} \\ a_{41} & a_{44} \end{vmatrix}
\begin{vmatrix} a_{22} & a_{23} \\ a_{42} & a_{43} \end{vmatrix}
+ \begin{vmatrix} a_{11} & a_{14} \\ a_{41} & a_{44} \end{vmatrix}
\begin{vmatrix} a_{12} & a_{13} \\ a_{42} & a_{43} \end{vmatrix}
+ \begin{vmatrix} a_{21} & a_{24} \\ a_{31} & a_{34} \end{vmatrix}
\begin{vmatrix} a_{22} & a_{32} \\ a_{23} & a_{33} \end{vmatrix}
$$

$$
= -\cos\theta + \begin{vmatrix} \langle v_1, v_2 \rangle & \langle v_1, v_3 \rangle \\ \langle v_4, v_2 \rangle & \langle v_4, v_3 \rangle \end{vmatrix} = -\cos\theta. \qquad \square
$$

Example 5.1. For each $\theta \in [0, \pi]$, we define a 1-parameter family of curves $\gamma_\theta : S^1 \to \mathbf{R}^4 \subset \operatorname{Im}\mathfrak{C}$ of helices of $\operatorname{Im}\mathfrak{C}$ by

$$
\gamma_\theta(s) = \frac{1}{35\sqrt{2}}\Big\{ -\sin\theta\sin(35s)k
$$

$$
+ \big(\cos\theta\sin(35s) + \cos(35s)i + 7\sin(5s)j + 7\cos(5s)k\big)\varepsilon \Big\}.
$$

The curvatures of γ_θ are $k_1 = 25$, $k_2 = 24$, $k_3 = 7$. On the other hand, its G_2-invariants are given by

$$k_1 = 25, \qquad\qquad \kappa_2(s) = \frac{12}{5}\sqrt{100 - 98\sin^2\theta\cos^2(5s)},$$

$$\rho_1(s) = -\frac{84\sqrt{2}}{5}\sin\theta\cos(5s), \qquad \rho_3(s) = \frac{35\sqrt{2}\sin\theta\cos(5s)}{50 - 49\sin^2\theta\cos^2(5s)},$$

$$\alpha(s) = -\frac{175\sqrt{2}\sin\theta\sin(5s)}{50 - 49\sin^2\theta\cos^2(5s)}, \quad \beta_1(s) = -\frac{350\cos\theta}{50 - 49\sin^2\theta\cos^2(5s)}.$$

We note that these G_2-invariants are not constant. Therefore if $\theta \not\equiv \tilde{\theta}$ (mod. 2π) then γ_θ and $\gamma_{\tilde{\theta}}$ are not G_2-congruent, but are $SO(7)$-congruent to each other. The (complicated) G_2-frame field along γ_θ is as follows:

$$\begin{aligned}
e_4 = \frac{1}{\sqrt{2}}\Big\{&-\sin\theta\cos(35s)k \\
&+ \big(\cos\theta\cos(35s) - \sin(35s)i + \cos(5s)j - \sin(5s)k\,\big)\varepsilon\Big\},
\end{aligned}$$

$$\begin{aligned}
e_1 = \frac{1}{5\sqrt{2}}\Big\{&7\sin\theta\sin(35s)k \\
&- \big(7\cos\theta\sin(35s) + 7\cos(35s)i + \sin(5s)j + \cos(5s)k\,\big)\varepsilon\Big\},
\end{aligned}$$

$$\begin{aligned}
e_5 = \frac{1}{10}\Big\{&(7\cos\theta - 1)i \\
&+ \big((\cos\theta+7)\cos(35s)\sin(5s) - (7\cos\theta+1)\sin(35s)\cos(5s)\big)j \\
&+ \big((\cos\theta+7)\cos(35s)\cos(5s) + (7\cos\theta+1)\sin(35s)\sin(5s)\big)k \\
&+ \sin\theta\Big(7\sin(35s)\sin(5s) + \cos(35s)\cos(5s) \\
&\qquad\quad + \big(7\sin(35s)\cos(5s) - \cos(35s)\sin(5s)\big)i + 7j\Big)\varepsilon\Big\},
\end{aligned}$$

$$\begin{aligned}
e_2 = \frac{12}{25\sqrt{2}\kappa_2}\Big\{&-7\sin\theta\cos(5s)(7\cos\theta - 1)i \\
&- 7\sin\theta\cos(5s)\big((\cos\theta + 7)\cos(35s)\sin(5s) \\
&\qquad\qquad\qquad - (7\cos\theta + 1)\sin(35s)\cos(5s)\,\big)j \\
&+ \sin\theta\big(50\cos(35s) - 7\cos(5s)(\cos\theta + 7)\cos(35s)\sin(5s) \\
&\qquad\qquad - 7\cos(5s)(7\cos\theta + 1)\sin(35s)\cos(5s)\big)k \\
&+ \big\{\big(-50\cos\theta\cos(35s) - 49\sin^2\theta\cos(5s)\sin(35s)\sin(5s) \\
&\qquad\qquad - 7\sin^2\theta\cos(5s)\cos(35s)\cos(5s)\big) \\
&\qquad + \big(50\sin(35s) - 49\sin^2\theta\cos(5s)\sin(35s)\cos(5s) \\
&\qquad\qquad + 7\sin^2\theta\cos(5s)\cos(35s)\sin(5s)\big)i \\
&\qquad + (50 - 49\sin^2\theta)\cos(5s)j - 50\sin(5s)k\big\}\varepsilon\Big\},
\end{aligned}$$

$$e_3 = \frac{12}{5\kappa_2}\Big\{-(7\cos\theta+1)i$$
$$+\big((7-\cos\theta)\cos(35s)\sin(5s)-(7\cos\theta-1)\sin(35s)\cos(5s)\big)j$$
$$+\big(\cos\theta(7-\cos\theta)\cos(35s)\cos(5s)+(7\cos\theta-1)\sin(35s)\sin(5s)\big)k$$
$$+\sin\theta\Big(7\sin(35s)\sin(5s)+(7\cos\theta-1)\cos(35s)\cos(5s)$$
$$+\cos(35s)\sin(5s)i-7\sin^2(5s)j-7\sin(5s)\cos(5s))k\Big)\varepsilon\Big\},$$

$$e_6 = \frac{12}{25\kappa_2}\Big\{-50\big(\cos\theta\cos(35s)\cos(5s)+\sin(35s)\sin(5s)\big)j$$
$$+\big(50\cos\theta\cos(35s)\sin(5s)-(50-49\sin^2\theta)\sin(35s)\cos(5s)\big)k$$
$$+\sin\theta\Big(50\cos(35s)\sin(5s)-49\cos\theta\sin(35s)\cos(5s)$$
$$+\cos(35s)\cos(5s)i-7\cos(35s)\sin(5s)j-7\cos^2(5s)k\Big)\varepsilon\Big\},$$

$$e_7 = \frac{24}{5\sqrt{2}\kappa_2}\Big\{-7\sin\theta\sin(5s)i-\sin\theta\cos(35s)j$$
$$+\Big(-\sin(35s)-\cos\theta\cos(35s)i+7\cos\theta\sin(5s)j$$
$$+7\cos\theta\cos(5s)k\Big)\varepsilon\Big\}.$$

Example 5.2. For each $\theta \in [0, \pi]$, we define a 1-parameter family $\gamma_\theta :$ $S^1 \to \mathbf{R}^4 \subset \mathrm{Im}\,\mathfrak{C}$ of helices in $\mathrm{Im}\,\mathfrak{C}$ by

$$\gamma_\theta(t) = r_1\cos(p_1 t)(-\sin\theta\,k+\cos\theta\,\varepsilon)$$
$$+r_1\sin(p_1 t)i\varepsilon+r_2\cos(p_2 t)j\varepsilon+r_2\sin(p_2 t)k\varepsilon,$$

where r_1, $r_2 > 0$, $p_1 p_2 \neq 0$, $p_1{}^2 \neq p_2{}^2$. If we set

$$c_1 = \sqrt{r_1{}^2 p_1{}^2+r_2{}^2 p_2{}^2}, \quad c_2 = \sqrt{r_1{}^2 p_1{}^4+r_2{}^2 p_2{}^4},$$

then its curvatures are

$$k_1 = \frac{c_2}{c_1{}^2}, \quad k_2 = \frac{r_1 r_2 p_1 p_2(p_1+p_2)(p_1-p_2)}{c_1{}^2 c_2}, \quad k_3 = \frac{p_1 p_2}{c_2}.$$

These are constants. On the other hand, its G_2-invariants are as follows:

$$k_1 = \frac{c_2}{c_1{}^2},$$
$$\kappa_2(t) = \frac{r_1 r_2 p_1 p_2(p_1+p_2)(p_1-p_2)}{c_1{}^2 c_2{}^2}\sqrt{c_2{}^2-r_1{}^2 p_1{}^4\sin^2\theta\sin^2(p_2 t)},$$
$$\rho_1(t) = -\frac{r_1{}^2 r_2{}^2 p_1 p_2(p_1+p_2)(p_1-p_2)}{c_1{}^2 c_2{}^2}\sin\theta\sin(p_2 t),$$

$$\rho_3(t) = \frac{r_2 p_1 p_2{}^3 \sin\theta \sin(p_2 t)}{c_2{}^2 - r_1{}^2 p_1{}^4 \sin^2\theta \sin^2(p_2 t)},$$

$$\alpha(t) = \frac{r_2 p_1 p_2{}^2 \sin\theta \cos(p_2 t)}{c_1(c_2{}^2 - r_1{}^2 p_1{}^4 \sin^2\theta \sin^2(p_2 t))},$$

$$\beta_1(t) = \frac{r_2 p_1 p_2 \cos\theta}{c_2{}^2 - r_1{}^2 p_1{}^4 \sin^2\theta \sin^2(p_2 t)}.$$

Therefore if $\theta \not\equiv \tilde{\theta}$ (mod. 2π) then γ_θ and $\gamma_{\tilde{\theta}}$ are not G_2-congruent, but are $SO(7)$-congruent congruent to each other. In particular, $\theta \equiv 0$ (mod. π), then the complete G_2-invariants are constant.

References

1. R. L. Bryant, *Submanifolds and special structures on the octonions*, J. Diff. Geom., 17 (1982) 185–232.
2. R. Harvey and H. B. Lawson, *Calibrated geometries*, Acta Math., 148 (1982) 47–157.
3. H. Hashimoto and K. Mashimo, *On some 3-dimensional CR submanifolds in S^6*, Nagoya Math. J. 156 (1999), 171-185.
4. H. Hashimoto and M. Ohashi, *Orthogonal almost complex structures of hypersurfaces of purely imaginary octonions*, Hokkaido Math. J. 39(2010), 351–387.
5. H. Hashimoto and M. Ohashi, *On generalized cylindrical helices and Lagrangian surfaces of $\boldsymbol{R}^4$*, in preparation.
6. B. O'Neill, *Elementary differential geometry*, Revised second edition. Elsevier/Academic Press, Amsterdam, 2006.
7. J. Oprea, *Differential geometry and its applications*, Second edition. Classroom Resource Materiers Series. Mathematical Associaton of America, Washington, DC, 2007.

Received January 17, 2011
Revised February 17, 2011

Proceedings of the 2nd International
Colloquium on Differential Geometry
and its Related Fields
Veliko Tarnovo, September 6–10, 2010

MAGNETIC JACOBI FIELDS
FOR KÄHLER MAGNETIC FIELDS

Toshiaki ADACHI*

*Department of Mathematics, Nagoya Institute of Technology,
Nagoya 466-8555, Japan
E-mail: adachi@nitech.ac.jp*

In this paper we improve the comparison theorem on magnetic Jacobi fields
for Kähler magnetic fields under an assumption that sectional curvatures are
bounded from above or from below by some constants.

Keywords: Kähler magnetic fields, Trajectories, Magnetic exponential maps,
Magnetic Jacobi fields, Comparison theorem.

1. Introduction

As a generalization of static magnetic fields on a Euclidean 3-space, a closed
2-form is said to be a magnetic field (see [7] for example). On a Kähler
manifold M with complex structure J we can consider natural uniform
magnetic fields which are given by constant multiples of the Kähler form $\mathbb{B}_J$.
We call them *Kähler magnetic fields*. For a Kähler magnetic field $\mathbb{B}_\kappa = \kappa \mathbb{B}_J$
with a constant κ, a smooth curve γ parameterized by its arc-length is said
to be a *trajectory* if it satisfies the differential equation $\nabla_{\dot\gamma}\dot\gamma = \kappa J\dot\gamma$. This
shows a motion of a unit charged particle with unit speed under the effect
of $\mathbb{B}_\kappa$. Trajectories for the trivial Kähler magnetic field $\mathbb{B}_0$ are geodesics. It
is needless to say that geodesics play a quite important role in the study
of Riemannian manifolds. Since trajectories for Kähler magnetic fields are
closely related with complex structure, the author hopes that they show
some properties of Kähler manifolds.

In this paper, we provide a fundamental result on variations of trajectories for Kähler magnetic fields. In order to connect tangent spaces and
manifolds we consider exponential maps induced by geodesics. As a fundamental property of geodesics we have the Hopf-Rinow theorem: A Rie-

*The author is partially supported by Grant-in-Aid for Scientific Research (C)
(No. 20540071) Japan Society of Promotion Science.

mannian manifold is complete if and only if the domain of the exponential map at a point coincides with the tangent space at that point. For given a magnetic field we can define corresponding maps induced by trajectories, which are called magnetic exponential maps. We are hence interested in their properties as a first step to study Kähler manifolds from the curve-theoretic point of view.

Since differentials of magnetic exponential maps corresponds to variations of trajectories, we need to study magnetic Jacobi fields which are induced by these variations. In [2] we gave a comparison theorem on magnetic Jacobi fields which corresponds to the Rauch's comparison theorem. Unfortunately, we need to pose a condition on sectional curvatures to get estimates on norms of magnetic Jacobi fields. Since the model spaces admitting Kähler magnetic fields are complex space forms, which are not of constant sectional curvatures, if we simply compare magnetic Jacobi fields on given two Kähler manifolds, we have a big loss in estimating their norms and estimates are not sharp. In this paper, we improve the proof and give a bit more sharp estimates under an assumption that sectional curvatures are bounded from above or from below by some constants.

2. Magnetic exponential maps

Let M be a complete Kähler manifold with complex structure J and Riemannian metric $\langle \ , \ \rangle$. If a smooth curve γ on M satisfies $\nabla_{\dot\gamma}\dot\gamma(t) = g(t)J\dot\gamma(t)$ with a function $g : \mathbb{R} \to \mathbb{R}$, we see $\frac{d}{dt}(\|\dot\gamma(t)\|^2) = 2g(t)\langle\dot\gamma(t), J\dot\gamma(t)\rangle = 0$, hence it has constant speed and is defined on whole $\mathbb{R}$. In particular, the domain of each trajectory for a Kähler magnetic field $\mathbb{B}_\kappa$ on M is $\mathbb{R}$.

For a unit tangent vector $v \in UM$ we denote by $\gamma_v = \gamma_{v;\kappa}$ a trajectory for $\mathbb{B}_\kappa$ with initial vector v. At a point $p \in M$, we define a *magnetic exponential map* $\mathbb{B}_\kappa\exp_p : T_pM \to M$ for $\mathbb{B}_\kappa$ of the tangent space T_pM by

$$\mathbb{B}_\kappa\exp_p(w) = \begin{cases} \gamma_{w/\|w\|}(\|w\|), & \text{if } w \neq 0_p, \\ p, & \text{if } w = 0_p. \end{cases}$$

Clearly, it is a usual exponential map defined by geodesics when $\kappa = 0$. Though we can consider trajectories as perturbed objects of geodesics, some of their properties are different from those of geodesics. On a complete Riemannian manifold, it is well-known that for given two distinct points there is a geodesic segment joining them, hence the image of an exponential map covers the base manifold (see [4, 6] for example). On the other hand, for magnetic exponential maps such a property does not hold in general.

To get the situation more clearly we here consider trajectories on a complex Euclidean space $\mathbb{C}^n$. On $\mathbb{C}^n$, the equation of a trajectory for $\mathbb{B}_\kappa$ is $\gamma'' = \sqrt{-1}\kappa\gamma'$, hence is a circle of radius $1/|\kappa|$ in the sense of Euclidean geometry. Thus they are closed of length $2\pi/|\kappa|$. If the distance of two points is greater than $\pi/|\kappa|$, then there are no trajectories for $\mathbb{B}_\kappa$ joining them. The image of a magnetic exponential map is therefore a closed ball of radius $\pi/|\kappa|$. Here, we show figures of variations of trajectories on $\mathbb{C}^n$. Each trajectory lies on some totally geodesic complex line $\mathbb{C}$. On $\mathbb{C}^n$ ($n \geq 2$) we can consider two kinds of variations; a variation of trajectories which lie on a same $\mathbb{C}$ and a variation of trajectories whose supporting complex lines are also varied.

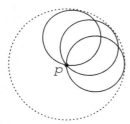

Fig. 1. variation to the complex direction on $\mathbb{C}^n$

Fig. 2. variation to a totally real direction on $\mathbb{C}^n$

3. Magnetic Jacobi fields

In order to study differentials of magnetic exponential maps, we need to investigate variations of trajectories. For a trajectory γ for $\mathbb{B}_\kappa$ on a Kähler manifold M, we call a smooth map $\alpha : (-\epsilon, \epsilon) \times \mathbb{R} \to M$ a variation of trajectories for $\mathbb{B}_\kappa$ if it satisfies

i) $\alpha_\gamma(0, t) = \gamma(t)$,
ii) for each s, the curve $t \mapsto \alpha(s, t)$ is a trajectory for $\mathbb{B}_\kappa$.

Just like variations of geodesics, this induces a vector field $\partial\alpha/\partial s|_{s=0}$ along γ. A vector field Y along a trajectory γ for $\mathbb{B}_\kappa$ is said to be a normal *magnetic Jacobi field* if it satisfies the following equations:

$$\begin{cases} \nabla_{\dot\gamma}\nabla_{\dot\gamma}Y - \kappa J\nabla_{\dot\gamma}Y + R(Y, \dot\gamma)\dot\gamma = 0, \\ \langle \nabla_{\dot\gamma}Y, \dot\gamma \rangle = 0. \end{cases} \tag{1}$$

Here, we should note that we suppose trajectories are parameterized by their arclengths. The second equality comes from the speed condition on trajectories. Trivially, $\dot{\gamma}$ is a normal magnetic Jacobi field along γ. Generally, if a vector field Z along γ satisfies $\nabla_{\dot{\gamma}}\nabla_{\dot{\gamma}}Z - \kappa J\nabla_{\dot{\gamma}}Z + R(Z,\dot{\gamma})\dot{\gamma} = 0$, then we see $\langle \nabla_{\dot{\gamma}}Z, \dot{\gamma}\rangle$ is constant. One may accept this additional condition. We can easily see that every normal magnetic Jacobi field for $\mathbb{B}_\kappa$ is obtained by a variation of trajectories for $\mathbb{B}_\kappa$ and vice versa (see Lemmas 1.3 and 1.5 in [2]).

For a vector field X along γ, we denote by $X = f_X\dot{\gamma} + g_X J\dot{\gamma} + X^\perp$ with functions f_X, g_X and a vector field $X^\perp$ which is orthogonal to both $\dot{\gamma}$ and $J\dot{\gamma}$ at each point. We put $X^\sharp = g_X J\dot{\gamma} + X^\perp$. By this expression the equations (1) turn to the following equations:

$$\begin{cases} f_Y' = \kappa g_Y, \\ (g_Y'' + \kappa^2 g_Y)J\dot{\gamma} + \nabla_{\dot{\gamma}}\nabla_{\dot{\gamma}}Y^\perp - \kappa J(\nabla_{\dot{\gamma}}Y^\perp) + R(Y^\sharp, \dot{\gamma})\dot{\gamma} = 0. \end{cases}$$

It is clear that for an arbitrary positive integer m we have $\langle \nabla_{\dot{\gamma}}^m Y^\perp, \dot{\gamma}\rangle = \langle \nabla_{\dot{\gamma}}^m Y^\perp, J\dot{\gamma}\rangle = 0$. Thus the second equality turns to

$$\begin{cases} g_Y'' + \kappa^2 g_Y + \langle R(Y^\sharp, \dot{\gamma})\dot{\gamma}, J\dot{\gamma}\rangle = 0, \\ \nabla_{\dot{\gamma}}\nabla_{\dot{\gamma}}Y^\perp - \kappa J(\nabla_{\dot{\gamma}}Y^\perp) + (R(Y^\sharp, \dot{\gamma})\dot{\gamma})^\perp = 0, \end{cases}$$

but we cannot treat $g_Y J\dot{\gamma}$ and $Y^\perp$ separately, in general.

We say a point $\gamma(t_0)$ to be a *magnetic conjugate point* for $p = \gamma(0)$ along γ if there is a non-trivial normal magnetic Jacobi field Y along γ satisfying $Y(0) = 0$ and $Y^\sharp(t_0) = 0$. As we can see by Figure 1, being different from variations of geodesics, at magnetic conjugate points trajectories may move. We call t_0 a magnetic conjugate value of p along γ. We denote by $c_\gamma(p)$ the minimum positive magnetic conjugate value of p along γ if magnetic conjugate points exist. When there are no magnetic conjugate points of p along γ we set $c_\gamma(p) = \infty$. By definition we have that $d(\mathbb{B}_\kappa \exp_p(t_0 v))$: $T_{t_0 v}(T_p M) \to T_{\gamma_v(t_0)}M$ is singular if and only if t_0 is a magnetic conjugate value of p along a trajectory γ_v for $\mathbb{B}_\kappa$.

4. Magnetic conjugate points on complex space forms

We here study magnetic conjugate values on a complex space form $\mathbb{C}M^n(c)$ of constant holomorphic sectional curvature c, which is one of a complex projective space $\mathbb{C}P^n(c)$, a complex Euclidean space $\mathbb{C}^n$ or a complex hyperbolic space $\mathbb{C}H^n(c)$ according to c is positive, zero or negative.

Example 4.1. On a complex Euclidean space $\mathbb{C}^n$, a normal magnetic Jacobi field Y along a trajectory γ for $\mathbb{B}_\kappa$ satisfying $Y(0) = 0$ is of the form

$$Y(t) = a\{(1 - \cos\kappa t)\dot{\gamma}(t) + \sin\kappa t J\dot{\gamma}(t)\} + (\gamma(t), A(1 - e^{\sqrt{-1}\kappa t}))$$

with a constant a and $A \in \mathbb{C}^n$ satisfying $A \perp \dot{\gamma}(0), J\dot{\gamma}(0)$.

Example 4.2. On a complex projective space $\mathbb{C}P^n(c)$, a normal magnetic Jacobi field Y along a trajectory γ for $\mathbb{B}_\kappa$ satisfying $Y(0) = 0$ is of the form

$$Y(t) = a\left\{\frac{\kappa}{\sqrt{\kappa^2+c}}(1 - \cos\sqrt{\kappa^2+c}\,t)\dot{\gamma}(t) + \sin\sqrt{\kappa^2+c}\,t\,J\dot{\gamma}(t)\right\}$$
$$+ d\varpi\left(\tilde{\gamma}(t), Ae^{\sqrt{-1}\kappa t/2}\sin\sqrt{\kappa^2+c}\,t/2\right),$$

where $\tilde{\gamma}$ is a horizontal lift of γ with respect to a Hopf fibration $\varpi : S^{2n+1} \to \mathbb{C}P^n$, a is a constant and $A \in \mathbb{C}^n$ satisfies $A \perp \tilde{\gamma}(0), J\tilde{\gamma}(0)$ and $A \perp \dot{\tilde{\gamma}}(0), J\dot{\tilde{\gamma}}(0)$.

Example 4.3. On a complex hyperbolic space $\mathbb{C}H^n(c)$, a normal magnetic Jacobi field Y along a trajectory γ for $\mathbb{B}_\kappa$ satisfying $Y(0) = 0$ is of the form

$$Y(t) = \begin{cases} a\left\{\dfrac{\kappa}{\sqrt{|c|-\kappa^2}}\left(\cosh\sqrt{|c|-\kappa^2}\,t - 1\right)\dot{\gamma}(t) + \sinh\sqrt{|c|-\kappa^2}\,t\,J\dot{\gamma}(t)\right\} \\ \qquad + d\varpi\left(\tilde{\gamma}(t), Ae^{\sqrt{-1}\kappa t/2}\sinh\sqrt{|c|-\kappa^2}\,t/2\right), \quad \text{if } |\kappa| < \sqrt{|c|}, \\[2mm] a\left\{\dfrac{|c|t^2}{2}\dot{\gamma}(t) + \kappa t\,J\dot{\gamma}(t)\right\} + d\varpi(\tilde{\gamma}(t), Ate^{\sqrt{-1}\kappa t/2}), \quad \text{if } \kappa = \pm\sqrt{|c|}, \\[2mm] a\left\{\dfrac{\kappa}{\sqrt{\kappa^2+c}}\left(1 - \cos\sqrt{\kappa^2+c}\,t\right)\dot{\gamma}(t) + \sin\sqrt{\kappa^2+c}\,t\,J\dot{\gamma}(t)\right\} \\ \qquad + d\varpi\left(\tilde{\gamma}(t), Ae^{\sqrt{-1}\kappa t/2}\sin\sqrt{\kappa^2+c}\,t/2\right), \quad \text{if } |\kappa| > \sqrt{|c|}, \end{cases}$$

where $\tilde{\gamma}$ is a horizontal lift of γ with respect to a fibration $\varpi : H_1^{2n+1} \to \mathbb{C}H^n$ of an anti-de Sitter space, a is a constant and $A \in \mathbb{C}^n$ satisfies $A \perp \tilde{\gamma}(0), J\tilde{\gamma}(0)$ and $A \perp \dot{\tilde{\gamma}}(0), J\dot{\tilde{\gamma}}(0)$.

For a constants κ and c we define two functions $\mathfrak{s}_\kappa(\cdot\,;c), \mathfrak{t}_\kappa(\cdot\,;c) : [0, \pi/\sqrt{\kappa^2+c}) \to [0, \infty)$ by

$$\mathfrak{s}_\kappa(t;c) = \begin{cases} (1/\sqrt{\kappa^2+c})\sin\sqrt{\kappa^2+c}\,t, & \text{if } \kappa^2+c > 0, \\ t, & \text{if } \kappa^2+c = 0, \\ (1/\sqrt{|c|-\kappa^2})\sinh\sqrt{|c|-\kappa^2}\,t, & \text{if } \kappa^2+c < 0, \end{cases}$$

$$\mathfrak{t}_\kappa(t;c) = \begin{cases} \sqrt{\kappa^2+c}\,\cot\sqrt{\kappa^2+c}\,t, & \text{if } \kappa^2+c > 0, \\ 1/t, & \text{if } \kappa^2+c = 0, \\ \sqrt{|c|-\kappa^2}\,\coth\sqrt{|c|-\kappa^2}\,t, & \text{if } \kappa^2+c < 0, \end{cases}$$

where we regard $1/\sqrt{\kappa^2+c}$ as infinity when $\kappa^2+c \le 0$. We here note that the case $\kappa^2+c < 0$ only occurs when $c < 0$. One can easily see the following by the definitions of these functions.

Lemma 4.1. *These functions* $\mathfrak{s}_\kappa(\cdot;c)$ *and* $\mathfrak{t}_\kappa(\cdot;c)$ *satisfy the following properties for* $0 < t < \pi/\sqrt{\kappa^2+c}$:

(1) *If* $|\kappa_1| < |\kappa_2|$, *then* $\mathfrak{t}_{\kappa_1}(t;c) > \mathfrak{t}_{\kappa_2}(t;c)$;
(2) $\mathfrak{s}_\kappa(\cdot;c)$ *is strictly increasing (i.e.* $\mathfrak{s}_\kappa'(t;c) > 0$) *on* $(0, \pi/2\sqrt{\kappa^2+c}\,)$;
(3) $\mathfrak{s}_\kappa(t;c) < 2\mathfrak{s}_\kappa(t/2;c)$ *when* $\kappa^2+c > 0$,
$\mathfrak{s}_\kappa(t;c) > 2\mathfrak{s}_\kappa(t/2;c)$ *when* $\kappa^2+c < 0$;
(4) $2\mathfrak{t}_\kappa(t;c) < \mathfrak{t}_\kappa(t/2;c)$ *when* $\kappa^2+c > 0$,
$2\mathfrak{t}_\kappa(t;c) > \mathfrak{t}_\kappa(t/2;c)$ *when* $\kappa^2+c < 0$.

It is clear that on $\mathbb{C}M^n(c)$ two trajectories γ_1, γ_2 for $\mathbb{B}_\kappa$ are congruent to each other in the strong sense, that is, there is a holomorphic isometry φ of $\mathbb{C}M^n(c)$ satisfying $\gamma_2(t) = \varphi \circ \gamma_1(t)$. Therefore magnetic conjugate values on $\mathbb{C}M^n(c)$ do not depend on the choice of trajectories for $\mathbb{B}_\kappa$ and the choice of initial points. We hence denote this magnetic conjugate value by $c(\kappa;c)$.

Proposition 4.1. *On a complex space form* $\mathbb{C}M^n(c)$ *we have the following.*

(1) $c(\kappa;c) = \pi/\sqrt{\kappa^2+c}$, *which is regarded as infinity when* $\kappa^2+c \leq 0$.
(2) *A normal magnetic Jacobi field* Y *along a trajectory* γ *for* $\mathbb{B}_\kappa$ *with* $Y(0) = 0$ *satisfies*

   i) $|g_Y(t)| = |g_Y'(0)| \times \mathfrak{s}_\kappa(t;c)$, $\|Y^\perp(t)\| = \|\nabla_{\dot\gamma} Y^\perp(0)\| \times 2\mathfrak{s}_\kappa(t/2;c)$;

   ii) $g_Y'(t) = g_Y(t) \times \mathfrak{t}_\kappa(t;c)$,
    $\langle \nabla_{\dot\gamma} Y^\perp(t), Y^\perp(t) \rangle = \|Y^\perp(t)\|^2 \times (1/2)\mathfrak{t}_\kappa(t/2;c)$

for $0 \leq t < \pi/\sqrt{\kappa^2+c}$. *In particular, in case* $n \geq 2$ *the following estimates hold:*

1) *When* $\kappa^2+c > 0$, *we have for* $0 \leq t < \pi/\sqrt{\kappa^2+c}$,

$$\|\nabla_{\dot\gamma} Y^\sharp(0)\| \times \mathfrak{s}_\kappa(t;c) \leq \|Y^\sharp(t)\| \leq \|\nabla_{\dot\gamma} Y^\sharp(0)\| \times 2\mathfrak{s}_\kappa(t/2;c),$$

$$\mathfrak{t}_\kappa(t;c) \leq \frac{\langle \nabla_{\dot\gamma} Y^\sharp(t), Y^\sharp(t) \rangle}{\|Y^\sharp(t)\|^2} \leq \frac{1}{2}\,\mathfrak{t}_\kappa\left(\frac{t}{2};c\right);$$

2) *When* $\kappa^2+c = 0$, *we have for* $t \geq 0$,

$$\|Y^\sharp(t)\| = \|\nabla_{\dot\gamma} Y^\sharp(0)\|\, t, \qquad \frac{\langle \nabla_{\dot\gamma} Y^\sharp(t), Y^\sharp(t) \rangle}{\|Y^\sharp(t)\|^2} = \frac{1}{t};$$

3) *When* $\kappa^2+c < 0$, *we have for* $t \geq 0$,

$$\|\nabla_{\dot\gamma} Y^\sharp(0)\| \times 2\mathfrak{s}_\kappa(t/2;c) \leq \|Y^\sharp(t)\| \leq \|\nabla_{\dot\gamma} Y^\sharp(0)\| \times \mathfrak{s}_\kappa(t;c),$$

$$\frac{1}{2}\,\mathfrak{t}_\kappa\left(\frac{t}{2};c\right) \leq \frac{\langle \nabla_{\dot\gamma} Y^\sharp(t), Y^\sharp(t) \rangle}{\|Y^\sharp(t)\|^2} \leq \mathfrak{t}_\kappa(t;c).$$

Proof. By Examples 4.1, 4.2 and 4.3 we get the equalities in (2–i) and in (2–ii). As $Y^\sharp = gJ\dot\gamma + Y^\perp$, we see $\nabla_{\dot\gamma}Y^\sharp = -\kappa g\dot\gamma + g'J\dot\gamma + \nabla_{\dot\gamma}Y^\perp$. Since $Y(0) = 0$, we have $\nabla_{\dot\gamma}Y^\sharp(0) = g'(0)J\dot\gamma(0) + \nabla_{\dot\gamma}Y^\perp(0)$. Therefore we obtain the estimates on $\|Y^\sharp(t)\|^2 = \{g_Y(t)\}^2 + \|Y^\perp(t)\|^2$ by Lemma 4.1 (3). In order to show the estimates on $\|Y^\sharp(t)\|^{-2}\langle\nabla_{\dot\gamma}Y^\sharp(t), Y^\sharp(t)\rangle$ we use the following inequalities: For constants β, δ and positive constants B, D, we have

$$\min\left\{\frac{\beta}{B}, \frac{\delta}{D}\right\} \leq \frac{\beta + \delta}{B + D} \leq \max\left\{\frac{\beta}{B}, \frac{\delta}{D}\right\}.$$

Because B and D are positive, these are obtained by direct computation. Since we see $\langle\nabla_{\dot\gamma}Y^\sharp(t), Y^\sharp(t)\rangle = g_Y'(t)g_Y(t) + \langle\nabla_{\dot\gamma}Y^\perp(t), Y^\perp(t)\rangle$, by applying the above inequalities we get

$$\frac{g_Y'(t)g_Y(t) + \langle\nabla_{\dot\gamma}Y^\perp(t), Y^\perp(t)\rangle}{\{g_Y(t)\}^2 + \|Y^\perp(t)\|^2} \geq \min\left\{\frac{g_Y'(t)}{g_Y(t)}, \frac{\langle\nabla_{\dot\gamma}Y^\perp(t), Y^\perp(t)\rangle}{\|Y^\perp(t)\|^2}\right\},$$

$$\frac{g_Y'(t)g_Y(t) + \langle\nabla_{\dot\gamma}Y^\perp(t), Y^\perp(t)\rangle}{\{g_Y(t)\}^2 + \|Y^\perp(t)\|^2} \leq \max\left\{\frac{g_Y'(t)}{g_Y(t)}, \frac{\langle\nabla_{\dot\gamma}Y^\perp(t), Y^\perp(t)\rangle}{\|Y^\perp(t)\|^2}\right\}.$$

Therefore by Lemma 4.1 (4) we obtain the desired estimates. □

The above proposition guarantees that every magnetic exponential map $\mathbb{B}_\kappa\exp_p$ on a complex space form $\mathbb{C}M^n(c)$ does not have singular points in the open ball $\{v \in T_pM \mid \|v\| < \pi/\sqrt{\kappa^2+c}\}$ of radius $\pi/\sqrt{\kappa^2+c}$. In particular, when $\kappa^2+c \leq 0$, magnetic exponential maps do not have singular points.

5. Comparison theorems on magnetic Jacobi fields

We now study differentials of magnetic exponential maps on general Kähler manifolds. For a trajectory γ for $\mathbb{B}_\kappa$ we denote by $\mathcal{J}_\gamma^\kappa$ the vector space of all normal magnetic Jacobi fields for $\mathbb{B}_\kappa$ along γ. In [2] we showed the following comparison theorem on magnetic Jacobi fields.

Proposition 5.1. *Let γ and $\tilde\gamma$ be trajectories for $\mathbb{B}_\kappa$ on Kähler manifolds M and $\widetilde M$, respectively. Assume*

(a) $\min\{\langle R(v, \dot\gamma(t))\dot\gamma(t), v\rangle \mid \|v\| = 1, v \perp \dot\gamma(t)\}$
    $\geq \max\{\langle\widetilde R(\tilde v, \dot{\tilde\gamma}(t))\dot{\tilde\gamma}(t), \tilde v\rangle \mid \|\tilde v\| = 1, \tilde v \perp \dot{\tilde\gamma}(t)\}$
    for $0 < t < c_\gamma(\gamma(0))$,
(b) $\dim(M) \geq \dim(\widetilde M)$.

We then have the following:

(1) $c_{\tilde{\gamma}}(\tilde{\gamma}(0)) \geq c_{\gamma}(\gamma(0))$;

(2) $Y \in \mathcal{J}_{\gamma}^{\kappa}$ with $Y(0) = 0$ satisfies

$$\|Y^{\sharp}(t)\| \leq \max\left\{ \|\widetilde{Y}^{\sharp}(t)\| \ \left| \ \begin{matrix} \widetilde{Y} \in \mathcal{J}_{\tilde{\gamma}}^{\kappa}, \\ \widetilde{Y}(0) = 0, \ \|\nabla_{\dot{\tilde{\gamma}}}\widetilde{Y}^{\sharp}(0)\| = \|\nabla_{\dot{\gamma}}Y^{\sharp}(0)\| \end{matrix} \right. \right\}$$

for $0 < t < c_{\gamma}(\gamma(0))$;

(3) $\widetilde{Y} \in \mathcal{J}_{\tilde{\gamma}}^{\kappa}$ with $\widetilde{Y}(0) = 0$ satisfies

$$\|\widetilde{Y}^{\sharp}(t)\| \geq \min\left\{ \|Y^{\sharp}(t)\| \ \left| \ \begin{matrix} Y \in \mathcal{J}_{\gamma}^{\kappa}, \\ Y(0) = 0, \ \|\nabla_{\dot{\gamma}}Y(0)\| = \|\nabla_{\dot{\tilde{\gamma}}}\widetilde{Y}(0)\| \end{matrix} \right. \right\}$$

for $0 < t < c_{\gamma}(\gamma(0))$.

We here improve this result under the assumption that sectional curvatures of a Kähler manifold is bounded from above. For a trajectory γ for $\mathbb{B}_{\kappa}$ on a Kähler manifold M we denote by $\mathcal{X}_{\gamma}([0,T])$ the set of vector fields along a trajectory-segment $\gamma|_{[0,T]}$ which are orthogonal to $\dot{\gamma}$ at every t ($0 \leq t \leq T$). For $X = g_X J\dot{\gamma} + X^{\perp} \in \mathcal{X}_{\gamma}([0,T])$, we define its index $\mathcal{I}_T$ by

$$\mathcal{I}_T(X) = \int_0^T \left\{ g_X'^2 - \kappa^2 g_X^2 + \langle \nabla_{\dot{\gamma}}X^{\perp} - \kappa JX^{\perp}, \nabla_{\dot{\gamma}}X^{\perp} \rangle - \langle R(X,\dot{\gamma})\dot{\gamma}, X \rangle \right\} dt.$$

For a normal magnetic Jacobi field Y we have

$$\mathcal{I}_T(Y^{\sharp}) = \langle \nabla_{\dot{\gamma}}Y^{\sharp}(T), Y^{\sharp}(T) \rangle - \langle \nabla_{\dot{\gamma}}Y^{\sharp}(0), Y^{\sharp}(0) \rangle.$$

By the same argument as for the usual index along a geodesic, we find the following (see Lemma 2.1 in [2]).

Lemma 5.1. *Let Y be a normal magnetic Jacobi field with $Y(0) = 0$ along a trajectory γ for a Kähler magnetic field. When $0 < T < c_{\gamma}(\gamma(0))$, an arbitrary $X \in \mathcal{X}_{\gamma}([0,T])$ with $X(0) = 0$ and $X(T) = Y^{\sharp}(T)$ satisfies $\mathcal{I}_T(X) \geq \mathcal{I}_T(Y^{\sharp})$. The equality holds if and only if $X = Y^{\sharp}$.*

With the aid of this lemma we can then get the following comparison theorem on magnetic Jacobi fields.

Theorem 5.1. *Let γ be a trajectory for a non-trivial Kähler magnetic field $\mathbb{B}_{\kappa}$ on a Kähler manifold M whose sectional curvatures satisfy $\mathrm{Riem}^M \leq c$ with some constant c. We then have the following.*

(1) $c_{\gamma}(\gamma(0)) \geq \pi/\sqrt{\kappa^2 + c}$.

(2) *If Y is a normal magnetic Jacobi field along γ satisfying $Y(0) = 0$, then for $0 < t < \pi/\sqrt{\kappa^2 + c}$ the function $\|Y^{\sharp}(t)\|/\mathfrak{s}_{\kappa}(t;c)$ is monotone increasing and Y satisfies the following:*

(a) $\|Y^\sharp(t)\| \geq \|\nabla_{\dot\gamma} Y^\sharp(0)\| \, \mathfrak{s}_\kappa(t; c),$

(b) $\langle \nabla_{\dot\gamma} Y^\sharp(t), Y^\sharp(t) \rangle \geq \|Y^\sharp(t)\|^2 \, \mathfrak{t}_\kappa(t; c).$

Here we regard $\pi/\sqrt{\kappa^2 + c}$ *as infinity when* $\kappa^2 + c \leq 0.$

In (2a) *or in* (2b), *if the equality holds at some* t_0 *with* $0 < t_0 < \pi/\sqrt{\kappa^2 + c},$ *then the equality holds at every* t *with* $0 \leq t \leq t_0,$ *and the magnetic Jacobi field* Y *is of the form* $Y^\sharp(t) = \pm\|\nabla_{\dot\gamma} Y^\sharp(0)\| \, \mathfrak{s}_\kappa(t; c) J\dot\gamma(t)$ *and sectional curvatures satisfy* $\langle R(J\dot\gamma, \dot\gamma)\dot\gamma, J\dot\gamma \rangle \equiv c$ *for* $0 \leq t \leq t_0.$

Proof. When the complex dimension of M is n we take a complex space form $\widehat{M} = \mathbb{C}M^n(4c)$ and a trajectory $\hat\gamma$ for $\mathbb{B}_\kappa$ on $\widehat{M}$. Let $P_\gamma^t : T_{\gamma(t)}M \to T_{\gamma(0)}M$ and $\widehat{P}_{\hat\gamma}^t : T_{\gamma(0)}\widehat{M} \to T_{\gamma(t)}\widehat{M}$ be parallel transformations along γ and $\hat\gamma$, respectively, and let $I : T_{\gamma(0)}M \to T_{\gamma(0)}\widehat{M}$ be a holomorphic linear isometry satisfying $I(\dot\gamma(0)) = \dot{\hat\gamma}(0)$. We set $\widehat{X}(t) = \widehat{P}_{\hat\gamma}^t \circ I \circ P_\gamma^t(Y^\perp(t))$. We then have $\widehat{X}$ is a vector field along $\hat\gamma$ which satisfies $\widehat{X}^\perp = \widehat{X}$. By use of the curvature condition, we have

$$\mathcal{I}_T(Y^\sharp) = \int_0^T \left\{ g_Y'^2 - \kappa^2 g_Y^2 + \langle \nabla_{\dot\gamma} Y^\perp - \kappa J Y^\perp, \nabla_{\dot\gamma} Y^\perp \rangle - \langle R(Y, \dot\gamma)\dot\gamma, Y \rangle \right\} dt$$

$$\geq \int_0^T \left\{ g_Y'^2 - \kappa^2 g_Y^2 + \langle \nabla_{\dot\gamma} Y^\perp - \kappa J Y^\perp, \nabla_{\dot\gamma} Y^\perp \rangle - c(g_Y^2 + \|Y^\perp\|^2) \right\} dt$$

$$= \int_0^T \left(g_Y'^2 - \kappa^2 g_Y^2 - c g_Y^2 \right) dt$$

$$\quad + \int_0^T \left\{ \langle \nabla_{\dot{\hat\gamma}} \widehat{X} - \kappa J \widehat{X}_T, \nabla_{\dot{\hat\gamma}} \widehat{X} \rangle - c\|\widehat{X}\|^2 \right\} dt$$

$$= \mathcal{I}_T(g_Y J\dot{\tilde\gamma}) + \mathcal{I}_T(\widehat{X}),$$

where $\tilde\gamma$ denotes a trajectory for $\mathbb{B}_\kappa$ on $\mathbb{C}M^1(c)$. We take a normal magnetic Jacobi field $\tilde{f}\dot{\tilde\gamma} + \tilde{g}J\dot{\tilde\gamma}$ for $\mathbb{B}_\kappa$ along $\tilde\gamma$ satisfying $\tilde{g}(0) = 0$, $\tilde{g}(T) = g_Y(T)$, and a normal magnetic Jacobi field $\widehat{Y}$ for $\mathbb{B}_\kappa$ along $\hat\gamma$ satisfying $\widehat{Y} = \widehat{Y}^\perp$, $\widehat{Y}(0) = 0$ and $\widehat{Y}(T) = \widehat{X}(T)$. By Lemma 5.1 and by the inequality $\mathfrak{t}_{\kappa/2}(T; c) > \mathfrak{t}_\kappa(T; c)$, we obtain

$$\langle \nabla_{\dot\gamma} Y^\sharp(T), Y^\sharp(T) \rangle \geq \mathcal{I}_T(g_Y J\dot{\tilde\gamma}) + \mathcal{I}_T(\widehat{X}) \geq \mathcal{I}_T(\tilde{g}J\dot{\tilde\gamma}) + \mathcal{I}_T(\widehat{Y})$$

$$\geq |g_Y(T)|^2 \, \mathfrak{t}_\kappa(T; c) + \|\widehat{X}(T)\|^2 \times \frac{1}{2}\mathfrak{t}_\kappa(T/2; 4c)$$

$$= |g_Y(T)|^2 \, \mathfrak{t}_\kappa(T; c) + \|Y^\perp(T)\|^2 \, \mathfrak{t}_{\kappa/2}(T; c)$$

$$\geq \|Y^\sharp(T)\|^2 \, \mathfrak{t}_\kappa(T; c).$$

We now consider a function $h(t) = \|\nabla_{\dot\gamma} Y^\sharp(0)\|^2 \, \mathfrak{s}_\kappa^2(t; c)$. It satisfies $h(0) =$

$h'(0) = 0$ and $h''(0) = 2\|\nabla_{\dot\gamma} Y^\sharp(0)\|^2$. We therefore have the following by de l'Hôpital's rule:

$$\lim_{t\downarrow 0} \frac{\|Y^\sharp(t)\|^2}{h(t)} = \lim_{t\downarrow 0} \frac{2\langle \nabla_{\dot\gamma} Y^\sharp(t), Y^\sharp(t)\rangle}{h'(t)}$$

$$= 2 \lim_{t\downarrow 0} \frac{\langle \nabla_{\dot\gamma}\nabla_{\dot\gamma} Y^\sharp(t), Y^\sharp(t)\rangle + \|\nabla_{\dot\gamma} Y^\sharp(t)\|^2}{h''(t)} = 1.$$

As we have

$$\frac{d}{dt}\left(\frac{\|Y^\sharp(t)\|^2}{h(t)} \right)\bigg|_{t=T}$$

$$= \frac{1}{h(T)}\left(2\langle \nabla_{\dot\gamma} Y^\sharp(T), Y^\sharp(T)\rangle - \|Y^\sharp(T)\|^2 \frac{h'(T)}{h(T)} \right)$$

$$= \frac{2}{h(T)}\left(\langle \nabla_{\dot\gamma} Y^\sharp(T), Y^\sharp(T)\rangle - \|Y^\sharp(T)\|^2 \, \mathfrak{t}_\kappa(T; c) \right) \geq 0,$$

we get $\|Y^\sharp(t)\| \geq \|\nabla_{\dot\gamma} Y^\sharp(0)\| \, \mathfrak{s}_\kappa(t; c)$.

We now consider the case that equality holds in (2a) or in (2b). If the equality in (2a) holds at some t_0, we then have the equality in (2b) holds for $0 < t \leq t_0$. Thus we are enough to consider the case that the equality holds in (2b). Since we have $\mathfrak{t}_{\kappa/2}(T; c) > \mathfrak{t}_\kappa(T; c)$ for $0 < T < \pi/\sqrt{\kappa^2+c}$, we find $Y^\perp(T) = 0$ hence obtain $\widehat{Y} \equiv 0$. As we need $\mathcal{I}_T(\widehat{X}) = \mathcal{I}_T(\widehat{Y}) = 0$ and $\mathcal{I}_T(g_Y J\dot\gamma) = \mathcal{I}_T(\tilde{g} J\dot\gamma)$, we see $Y^\perp \equiv 0$ and $g_Y \equiv \tilde{g}$. Since we also need $\langle R(Y, \dot\gamma)\dot\gamma, Y\rangle = c(g_Y^2 + \|Y^\perp\|^2)$, we get the conclusion.   □

As a consequence of Theorem 5.1 we get the following property on magnetic exponential maps.

Corollary 5.1. *If sectional curvatures of a Kähler manifold M satisfy* $\mathrm{Riem}^M \leq c$, *then every magnetic exponential map* $\mathbb{B}_\kappa \exp_p$ *on M do not have singular points in the open ball* $\{v \in T_p M \mid \|v\| < \pi/\sqrt{\kappa^2+c}\}$. *In particular, when $\kappa^2+c \leq 0$, magnetic exponential maps do not have singular points.*

Next we give an estimate of magnetic Jacobi fields from above. We define a function $\mathfrak{u}_\kappa(t; c)$ by

$$\mathfrak{u}_\kappa(t; c) = \begin{cases} \dfrac{1 - \cos\sqrt{\kappa^2+c}\, t}{\kappa^2+c} = 2\mathfrak{s}_\kappa^2(t/2; c), & \text{if } \kappa^2+c > 0, \\[2ex] t^2/2, & \text{if } \kappa^2+c = 0, \\[2ex] \dfrac{\cosh\sqrt{|c|-\kappa^2}\, t - 1}{|c|-\kappa^2} = 2\mathfrak{s}_\kappa^2(t/2; c), & \text{if } \kappa^2+c < 0. \end{cases}$$

Theorem 5.2. *Let γ be a trajectory for a non-trivial Kähler magnetic field $\mathbb{B}_\kappa$ on a Kähler manifold M whose sectional curvatures satisfy $\mathrm{Riem}^M \geq c$ with some constant c. We then have the following.*

(1) $c_\gamma(\gamma(0)) \leq 2\pi/\sqrt{\kappa^2 + 4c}$.

(2) *If Y is a normal magnetic Jacobi field along γ with $Y(0) = 0$, then for $0 \leq t \leq c_\gamma(\gamma(0))$ the function $\mathfrak{s}_{\kappa/2}(t; c)/\|Y^\sharp(t)\|$ is monotone increasing and Y satisfies the following:*

    (a) $\|Y^\sharp(t)\| \leq \|\nabla_{\dot\gamma} Y^\sharp(0)\| \, \mathfrak{s}_{\kappa/2}(t; c)$,

    (b) $|f_Y(t)| \leq |\kappa| \, \|\nabla_{\dot\gamma} Y^\sharp(0)\| \, \mathfrak{u}_{\kappa/2}(t; c)$,

    (c) $\langle \nabla_{\dot\gamma} Y^\sharp(t), Y^\sharp(t) \rangle \leq \|Y^\sharp(t)\|^2 \, \mathfrak{t}_{\kappa/2}(t; c)$.

In (2a), in (2b) or in (2c), if the equality holds at some t_0 with $0 < t_0 < c_\gamma(\gamma(0))$, then the equality holds at every t with $0 \leq t \leq t_0$, and the magnetic Jacobi field is of the form

$$Y^\sharp(t) = Y^\perp(t)$$
$$= \|\nabla_{\dot\gamma} Y^\perp(0)\| \, \mathfrak{s}_{\kappa/2}(t; c)\big\{\cos(\kappa t/2)\, E(t) + \sin(\kappa t/2)\, JE(t)\big\}$$

with a parallel vector field E along γ satisfying the initial condition $E(0) = \nabla_{\dot\gamma} Y^\perp(0)/\|\nabla_{\dot\gamma} Y^\perp(0)\|$ and sectional curvatures satisfy $\langle R(F, \dot\gamma)\dot\gamma, F \rangle \equiv c$ for $0 \leq t \leq t_0$ with the vector field $F(t) = \cos(\kappa t/2)\, E(t) + \sin(\kappa t/2)\, JE(t)$ along γ.

Proof. We take a trajectory $\tilde\gamma$ for $\mathbb{B}_\kappa$ on $\mathbb{C}M^1(c)$ and a trajectory $\hat\gamma$ for $\mathbb{B}_\kappa$ on $\mathbb{C}M^n(4c)$. Borrowing notations in the proof of Theorem 5.1, for an arbitrary positive T with $T < c_\gamma(\gamma(0))$, we take a magnetic Jacobi field $\widehat{Y}_T$ for $\mathbb{B}_\kappa$ along $\hat\gamma$ which satisfies $\widehat{Y}_T(0) = 0$ and $\widehat{Y}_T(T) = \widehat{P}_{\hat\gamma}^T \circ I \circ P_\gamma^T\big(Y^\perp(T)\big)$. We also take a magnetic Jacobi field $\tilde g J\dot{\tilde\gamma}$ for $\mathbb{B}_\kappa$ along $\tilde\gamma$ satisfying $\tilde g(0) = 0$ and $\tilde g(T) = g_Y(T)$, We set a vector field along γ by $X_T(t) = \tilde g(t) J\dot\gamma(t) + \big(\widehat{P}_{\hat\gamma}^t \circ I \circ P_\gamma^t\big)^{-1}\big(\widehat{Y}_T(t)\big)$. We then have $X_T(T) = Y^\sharp(T)$, hence we obtain

$$\|Y^\sharp(T)\|^2 \, \mathfrak{t}_{\kappa/2}(T; c)$$
$$\geq g_Y(T)^2 \mathfrak{t}_\kappa(T; c) + \|Y^\perp(T)\|^2 \, \mathfrak{t}_{\kappa/2}(T; c)$$
$$= \mathcal{I}_T(\tilde g J\dot{\tilde\gamma}) + \mathcal{I}_T(\widehat{Y}_T)$$
$$= \int_0^T \Big\{ \tilde g'^2 - \kappa^2 \tilde g^2 + \langle \nabla_{\dot{\hat\gamma}} \widehat{Y}_T - \kappa J\widehat{Y}_T, \nabla_{\dot{\hat\gamma}} \widehat{Y}_T \rangle - c\big(\tilde g^2 + \|\widehat{Y}_T\|^2\big) \Big\} dt$$
$$\geq \int_0^T \Big\{ \tilde g'^2 - \kappa^2 \tilde g^2 + \langle \nabla_{\dot\gamma} X_T^\perp - \kappa J X_T^\perp, \nabla_{\dot\gamma} X_T^\perp \rangle - \langle R(X_T, \dot\gamma)\dot\gamma, X_T \rangle \Big\} dt$$
$$= \mathcal{I}_T(X_T) \geq \mathcal{I}_T(Y^\sharp) = \langle Y^\sharp(T), \nabla_{\dot\gamma} Y^\sharp(T) \rangle.$$

We consider a smooth function $h(t) = \|\nabla_{\dot{\gamma}} Y^{\sharp}(0)\|^2 \, \mathfrak{s}^2_{\kappa/2}(t; c)$. Along the same lines as in the proof of Theorem 5.1, we have $\lim_{t \downarrow 0} h(t)/\|Y^{\sharp}(t)\|^2 = 1$ and have

$$
\begin{aligned}
\frac{d}{dt}\left(\frac{h(t)}{\|Y^{\sharp}(t)\|^2}\right)\bigg|_{t=T} \\
&= \frac{h(T)}{\|Y^{\sharp}(T)\|^4}\left(\|Y^{\sharp}(T)\|^2 \frac{h'(T)}{h(T)} - 2\langle Y^{\sharp}(T), \nabla_{\dot{\gamma}} Y^{\sharp}(T)\rangle\right) \\
&= \frac{2h(T)}{\|Y^{\sharp}(T)\|^4}\left(\|Y^{\sharp}(T)\|^2 \, \mathfrak{t}_{\kappa/2}(T; c) - \langle Y^{\sharp}(T), \nabla_{\dot{\gamma}} Y^{\sharp}(T)\rangle\right) \geq 0,
\end{aligned}
$$

hence we get the estimate (2a). As we have $f'_Y(t) = \kappa g_Y(t)$, we have

$$
\begin{aligned}
|f_Y(t)| = \left|\int_0^t f'_Y(s)ds\right| &\leq \int_0^t |f'_Y(s)|\, ds = |\kappa| \int_0^t |g_Y(s)|\, ds \\
&\leq |\kappa| \int_0^t \|Y^{\sharp}(s)\|\, ds \leq |\kappa| \, \|\nabla_{\dot{\gamma}} Y^{\sharp}(0)\| \int_0^t \mathfrak{s}_{\kappa/2}(s; c)\, ds,
\end{aligned}
$$

hence get the estimate (2b).

If the equality in (2b) holds at some t_0, we then have the equality in (2a) holds for $0 < t \leq t_0$. If the equality in (2a) holds at some t_0, then we have the equality in (2c) holds for $0 < t \leq t_0$. Thus we are enough to consider the case that the equality holds in (2c). If the equality holds in (2c) at some t_0, then we find $X_{t_0} \equiv Y^{\sharp}$, $\langle R(X_{t_0}, \dot{\gamma})\dot{\gamma}, X_{t_0}\rangle \equiv c$, and find $g_Y(t_0) = 0$ because $\mathfrak{t}_{\kappa/2}(t_0; c) > \mathfrak{t}_{\kappa}(t_0; c)$. This shows $\tilde{g} \equiv 0$, hence guarantees that $Y^{\perp} \equiv Y^{\sharp}$ is a normal magnetic Jacobi field along γ and that $\langle R(Y^{\perp}, \dot{\gamma})\dot{\gamma}, Y^{\perp}\rangle \equiv c\|Y^{\perp}\|^2$. Since the normal magnetic Jacobi field $\widehat{Y}_{t_0}$ on $\mathbb{C}M^n(4c)$ satisfies $\widehat{Y}_{t_0} = \widehat{Y}^{\perp}_{t_0}$ and its expression is given in Examples 4.1, 4.2 and 4.3, we get the expression of $Y^{\perp} = X_{t_0}$. We hence obtain the conclusion. $\qquad \square$

As we mentioned before, Figures 1 and 2 suggest us that the meaning of variations of trajectories along a trajectory γ to the direction of $J\dot{\gamma}(0)$ and that to totally real directions should be different. We are hence interested in giving an estimate of $|g_Y|$ of a normal magnetic Jacobi field Y especially from below. Still we have no idea.

Under an assumption that sectional curvatures are bounded from above, we can give an estimate of norms of normal magnetic Jacobi fields. For this sake we need to study the meaning of a normal magnetic Jacobi field Y with $\nabla Y^{\sharp}(0) - (\kappa/2)JY^{\sharp}(0) = 0$ and $f_Y(0) = 0$, we will discuss about such an estimate somewhere else.

References

1. T. Adachi, *Kähler magnetic flows on a manifold of constant holomorphic sectional curvature*, Tokyo J. Math. 18(1995), 473–483.
2. _____, *A comparison theorem for magnetic Jacobi fields*, Proc. Edinburgh Math. Soc. 40(1997), 293–308.
3. _____, *A theorem of Hadamard-Cartan type for Kähler magnetic fields*, to appear in J. Math. Soc. Japan.
4. J. Cheeger and Ebin, *Comparison Theorems in Riemannian Geometry*, North-Holland Publ. Co., Amsterdam (1975).
5. N. Gouda, *Magnetic flows of Anosov type*, Tohoku Math. J. 49(1997), 165–183.
6. T. Sakai, *Riemannian Geometry*, Shokabo, Tokyo (1992) in Japanese; Transl. Math. Monographs 149, A.M.S. (1996).
7. T. Sunada, *Magnetic flows on a Riemann surface*, Proc. KAIST Math. Workshop 8(1993), 93–108.

Received January 5, 2011
Revised   March 24, 2011

Proceedings of the 2nd International
Colloquium on Differential Geometry
and its Related Fields
Veliko Tarnovo, September 6–10, 2010

GEOMETRY FOR q-EXPONENTIAL FAMILIES

Hiroshi MATSUZOE

*Department of Computer Science and Engineering,
Graduate School of Engineering, Nagoya Institute of Technology,
Nagoya, Aichi 466-8555 Japan
E-mail: matsuzoe@nitech.ac.jp*

Atsumi OHARA

*Department of Electrical and Electronics Engineering,
Graduate School of Engineering, University of Fukui,
Fukui 910-8507 Japan
E-mail: ohara@fuee.u-fukui.ac.jp*

Geometry for q-exponential families is studied in this paper. A q-exponential family is a set of probability distributions, which is a natural generalization of the standard exponential family. A q-exponential family has information geometric structure and a dually flat structure. To describe these relations, generalized conformal structures for statistical manifolds are studied in this paper. As an application of geometry for q-exponential families, a geometric generalization of statistical inference is also studied.

Keywords: q-exponential family, q-product, Information geometry, Tsallis statistics, Statistical manifold, Divergence.

Introduction

An exponential family is a set of probability distributions such as a set of normal distributions, of Poisson distributions, or of gamma distributions, etc. Such probability distributions decay exponentially. However, in complex systems, probability distributions often have long tails, that is, probability distributions do not decay exponentially. The q-normal distribution which is frequently discussed in Tsallis nonextensive statistical mechanics [18] is a typical example of such probability distributions.

In this paper, we consider q-exponential families. A q-exponential family is a natural generalization of the standard exponential family, and which includes the set of q-normal distributions. From the viewpoint of information geometry, it is known that an exponential family has a dually flat structure (see [1]). We will see that q-exponential families naturally have dually flat structures.

A q-exponential family also has information geometric structure, that is, a q-exponential family has the Fisher metric and α-connections. Hence a q-exponential family has two kinds of statistical manifold structures. Thus, we consider relations of these structures using generalized conformal equivalence relations on statistical manifolds.

In the later part of this paper, we consider statistical inferences for q-exponential families. Generalizations of independence or likelihood functions have been introduced in machine learning theory [4] or in Tsallis statistics [16]. We show that dually flat structures on q-exponential families work naturally for such generalized statistical inferences.

1. Preliminaries

In this section, we review geometry of statistical models and related geometry (cf.[1, 15]). We assume that all objects are smooth throughout this paper. We also assume that the manifold is simply connected since we will discuss geometry of statistical models.

1.1. *Statistical models*

Let $\mathcal{X}$ be a total sample space and let Ξ be an open domain of $\boldsymbol{R}^n$. We say that S is a *statistical model* or a *parametric model* on $\mathcal{X}$ if S is a set of probability densities with parameter $\xi \in \Xi$ such that

$$S = \left\{ p(x;\xi) \middle| \int_{\mathcal{X}} p(x;\xi)dx = 1, p(x;\xi) > 0, \xi \in \Xi \subset \boldsymbol{R}^n \right\}.$$

Under suitable conditions, S can be regarded as a manifold with a local coordinate system $\{\xi^1, \ldots, \xi^n\}$ (see [1]).

For a statistical model S, we define a function $g^F_{ij}(\xi) : \Xi \to \boldsymbol{R}$ by the following formula:

$$g^F_{ij}(\xi) := \int_{\mathcal{X}} \left(\frac{\partial}{\partial \xi^i} \log p(x;\xi) \right) \left(\frac{\partial}{\partial \xi^j} \log p(x;\xi) \right) p(x;\xi)dx$$

$$= E_\xi[\partial_i l_\xi \partial_j l_\xi].$$

Here, for simplicity, we used following notations:

$$E_\xi[f] = \int_{\mathcal{X}} f(x)p(x;\xi)dx, \qquad \text{(the expectation of } f(x) \text{ at } p(x;\xi)),$$

$$l_\xi = l(x;\xi) = \log p(x;\xi), \quad \text{(the log likelihood of } p(x;\xi)),$$

$$\partial_i = \frac{\partial}{\partial \xi^i}.$$

We assume that $g_{ij}^F(\xi)$ is finite for all i, j, ξ. Set a matrix $g^F = (g_{ij}^F)$, then we can check that g^F is symmetric and non-negative definite. We assume that g^F is positive definite. Then g^F is a Riemannian metric on S. We call g^F the *Fisher metric* on S.

For $\alpha \in \mathbf{R}$, we define the α-*connection* $\nabla^{(\alpha)}$ by the following formulas:

$$\Gamma_{ij,k}^{(\alpha)}(\xi) = E_\xi\left[\left(\partial_i\partial_j l_\xi + \frac{1-\alpha}{2}\partial_i l_\xi \partial_j l_\xi\right)(\partial_k l_\xi)\right],$$

$$h(\nabla_{\partial_i}^{(\alpha)}\partial_j, \partial_k) = \Gamma_{ij,k}^{(\alpha)}.$$

We can check that $\nabla^{(\alpha)}$ is torsion-free and $\nabla^{(0)}$ is the Levi-Civita connection of the Fisher metric. It is known that ± 1-connections are more important than the Levi-Civita connection in geometric theory of statistical inferences. We call $\nabla^{(1)}$ the *exponential connection* and $\nabla^{(-1)}$ the *mixture connection*.

For α-connections, the following formula holds

$$Xg^F(Y, Z) = g^F(\nabla_X^{(\alpha)}Y, Z) + g^F(Y, \nabla_X^{(-\alpha)}Z).$$

The connections $\nabla^{(\alpha)}$ and $\nabla^{(-\alpha)}$ are said to be *dual* (or *conjugate*) with respect to g^F. For arbitrary $\alpha, \beta \in \mathbf{R}$, the difference between the α-connection and the β-connection is given by

$$\Gamma_{ij,k}^{(\beta)} = \Gamma_{ij,k}^{(\alpha)} + \frac{\alpha - \beta}{2}C_{ijk}^F,$$

where

$$C_{ijk}^F(\xi) = E_\xi[\partial_i l_\xi \partial_j l_\xi \partial_k l_\xi].$$

The $(0, 3)$-tensor field C^F determined by C_{ijk}^F is called a *cubic form*. The covariant derivative of the Fisher metric g^F satisfies $(\nabla_X^{(\alpha)}g^F)(Y, Z) = \alpha C^F(X, Y, Z)$.

We say that a statistical model S is an *exponential family* if

$$S = \left\{ p(x; \theta) \;\middle|\; p(x; \theta) = \exp\left[Z(x) + \sum_{i=1}^{n}\theta^i F_i(x) - \psi(\theta)\right], \theta \in \Theta \subset \mathbf{R}^n \right\},$$

where Θ is a parameter space, $Z, F_1, \cdots, F_n$ are random variables on $\mathcal{X}$ and ψ is a function on Θ. The coordinate system $\{\theta^i\}$ is called the *natural parameters*.

Proposition 1.1. *For an exponential family S, the natural parameters $\{\theta^i\}$ is an affine coordinate system with respect to $\nabla^{(1)}$, that is, $\Gamma_{ij}^{(1)\,k} \equiv 0$ $(i, j, k = 1, \ldots, n)$, and the 1-connection $\nabla^{(1)}$ is flat.*

For simplicity, we set $Z = 0$. It is possible to assume this condition without loss of generality. We say that M is a *curved exponential family* of S if M is a submanifold of S such that

$$M = \{p(x; \theta(u)) \,|\, p(x; \theta(u)) \in S, \; u \in U \subset \mathbf{R}^m \} .$$

Example 1.1 (normal distributions). *Let S be the set of normal distributions,*

$$S = \left\{ p(x; \mu, \sigma) \;\middle|\; p(x; \mu, \sigma) = \frac{1}{\sqrt{2\pi}\sigma} \exp\left[-\frac{(x - \mu)^2}{2\sigma^2} \right] \right\} .$$

Here, the sample space $\mathcal{X}$ is $\mathbf{R}$, and the parameter space is the upper half plane $\Xi = \{(\mu, \sigma)\} |-\infty < \mu < \infty, 0 < \sigma < \infty\}$.

The Fisher metric in (μ, σ)-coordinate is given by

$$(g_{ij}^F) = \frac{1}{\sigma^2} \begin{pmatrix} 1 & 0 \\ 0 & 2 \end{pmatrix} .$$

Hence S is a space of constant negative curvature $-1/2$.

Let us change parameters as follows:

$$\theta^1 = \frac{\mu}{\sigma^2}, \quad \theta^2 = -\frac{1}{2\sigma^2} .$$

Set

$$Z(x) = 0, \; F_1(x) = x, \; F_2(x) = x^2,$$
$$\psi(\theta) = \frac{\mu^2}{2\sigma^2} + \log(\sqrt{2\pi}\sigma) = -\frac{(\theta^1)^2}{4\theta^2} + \frac{1}{2} \log\left(-\frac{\pi}{\theta^2} \right),$$

then we obtain

$$p(x; \mu, \sigma) = \frac{1}{\sqrt{2\pi}\sigma} \exp\left[-\frac{(x - u)^2}{2\sigma^2} \right]$$
$$= \exp\left[\frac{\mu}{\sigma^2}x - \frac{1}{2\sigma^2}x^2 - \frac{\mu^2}{2\sigma^2} - \log(\sqrt{2\pi}\sigma) \right]$$
$$= \exp\left[x\theta^1 + x^2\theta^2 - \psi(\theta) \right] .$$

This implies that the set of normal distributions is an exponential family.

For an exponential family, the Fisher metric and the cubic form in $\{\theta^i\}$-coordinate are given by

$$g_{ij}^F(\theta) = \partial_i \partial_j \psi(\theta), \tag{1}$$
$$C_{ijk}^F(\theta) = \partial_i \partial_j \partial_k \psi(\theta). \tag{2}$$

The *expectation parameters* $\{\eta_i\}$ are given by $\eta_i = E[F_i(x)]$, and $\{\eta_i\}$ is a $\nabla^{(-1)}$-affine coordinate system.

1.2. Statistical manifolds

Let (M, h) be a semi-Riemannian manifold, and let ∇ be a torsion-free affine connection on M. We sat that the triplet (M, ∇, h) is a *statistical manifold* if ∇h is a totally symmetric $(0, 3)$-tensor field. Obviously, a statistical model has many statistical manifold structures.

For a statistical manifold (M, ∇, h), we define the *dual connection* ∇^* with respect to h by

$$Xh(Y, Z) = h(\nabla_X Y, Z) + h(Y, \nabla_X^* Z).$$

The connection ∇^* is torsion-free and $\nabla^* h$ is also symmetric. Hence the triplet (M, ∇^*, h) is a statistical manifold. We call (M, ∇^*, h) the *dual statistical manifold* of (M, ∇, h).

Proposition 1.2. *Let (M, h) be a semi-Riemannian manifold and let C be a totally symmetric $(0, 3)$-tensor field. Denote by $\nabla^{(0)}$ the Levi-Civita connection $\nabla^{(0)}$ with respect to h. We define an affine connection $\nabla^{(\alpha)}$ by*

$$h(\nabla_X^{(\alpha)} Y, Z) := h(\nabla_X^{(0)} Y, Z) - \frac{\alpha}{2} C(X, Y, Z).$$

Then, the connections $\nabla^{(\alpha)}$ and $\nabla^{(-\alpha)}$ are torsion-free affine connections mutually dual with respect to h, and the covariant derivative $\nabla^{(\alpha)} h$ is totally symmetric. Hence $(M, \nabla^{(\alpha)}, h)$ and $(M, \nabla^{(-\alpha)}, h)$ are statistical manifolds.

The connection ∇ is flat if and only if ∇^* is flat. In this case, we say that (M, h, ∇, ∇^*) is a *dually flat space*. Since the connection ∇ is flat, there exists an affine coordinate system $\{\theta^i\}$ on M. In addition, there exits a ∇^*-affine coordinate system $\{\eta_i\}$ such that

$$h\left(\frac{\partial}{\partial \theta^i}, \frac{\partial}{\partial \eta_j}\right) = \delta_i^j.$$

We say that $\{\eta_i\}$ is the *dual coordinate system* of $\{\theta^i\}$ with respect to h.

Proposition 1.3. *Let (M, h, ∇, ∇^*) be a dually flat space. Suppose that $\{\theta^i\}$ is a ∇-affine coordinate system, and $\{\eta_i\}$ is the dual coordinate system of $\{\theta^i\}$. Then there exist functions ψ and ϕ on M such that*

$$\frac{\partial \psi}{\partial \theta^i} = \eta_i, \quad \frac{\partial \phi}{\partial \eta_i} = \theta^i, \quad \psi(p) + \phi(p) - \sum_{i=1}^{n} \theta^i(p)\eta_i(p) = 0. \tag{3}$$

In addition, the following formulas hold:

$$h_{ij} = \frac{\partial^2 \psi}{\partial \theta^i \partial \theta^j}, \quad h^{ij} = \frac{\partial^2 \phi}{\partial \eta_i \partial \eta_j}, \tag{4}$$

where (h_{ij}) *is the component matrix of a semi-Riemannian metric* h *with respect to* $\{\theta^i\}$, *and* (h^{ij}) *is the inverse matrix of* (h_{ij}).

The functions ψ and ϕ are called the θ-*potential* and the η-*potential*, respectively. The relation (3) is called the *Legendre transformation*. From Equation (4), the semi-Riemannian metric h is a Hessian metric. Hence we also say that (M, ∇, h) is a *Hessian manifold* [15].

Definition 1.1. We say that a function ρ on $M \times M$ is the *(canonical) divergence* on (M, h, ∇, ∇^*) if

$$\rho(p\|q) := \psi(p) + \phi(q) - \sum_{i=1}^{n} \theta^i(p)\eta_i(q), \quad (p, q \in M). \tag{5}$$

We remark that the definition of ρ is independent of the choice of affine coordinate system on M.

1.3. *Generalized conformal relations on statistical manifolds*

We give a brief summary of generalized conformal relations on statistical manifolds. Generalized conformal structures on statistical manifolds have been studied in affine differential geometry (see [5, 6, 7, 8]).

Definition 1.2. Suppose (M, ∇, h) and $(M, \bar{\nabla}, \bar{h})$ are statistical manifolds. We say that (M, ∇, h) and $(M, \bar{\nabla}, \bar{h})$ are *conformally-projectively equivalent* if there exist two functions κ and λ such that

$$\bar{h}(X, Y) = e^{\kappa + \lambda} h(X, Y),$$
$$\bar{\nabla}_X Y = \nabla_X Y - h(X, Y)\text{grad}_h\lambda + d\kappa(Y)\, X + d\kappa(X)\, Y,$$

where $\text{grad}_h\lambda$ is the gradient vector field of λ with respect to h.

In particular, for a constant $\alpha \in \mathbf{R}$, we say that two statistical manifolds are α-*conformally equivalent* if there exists a function λ on M such that

$$\bar{h}(X, Y) = e^{\lambda} h(X, Y),$$
$$\bar{\nabla}_X Y = \nabla_X Y - \frac{1+\alpha}{2} h(X, Y)\text{grad}_h\lambda + \frac{1-\alpha}{2}\left\{ d\lambda(Y)\, X + d\lambda(X)\, Y \right\}.$$

A statistical manifold (M, ∇, h) is called α-*conformally flat* if (M, ∇, h) is locally α-conformally equivalent to some flat statistical manifold.

We remark that the conformal-projective equivalence relation or the α-conformal equivalence relation are natural generalizations of conformal equivalence relation for Riemannian manifolds. In fact, suppose that (M, g) and $(M, \bar{g})$ are Riemannian manifolds, and $\nabla^{(0)}$ and $\bar{\nabla}^{(0)}$ denote their Levi-Civita connections. If g and $\bar{g}$ are conformally equivalent, then the following formulas fold.

$$\bar{g}(X, Y) = e^{2\lambda} g(X, Y),$$
$$\bar{\nabla}_X^{(0)} Y = \nabla_X^{(0)} Y - h(X, Y)\mathrm{grad}_h \lambda + d\lambda(Y)\, X + d\lambda(X)\, Y.$$

This implies that $(M, \nabla^{(0)}, g)$ and $(M, \bar{\nabla}^{(0)}, \bar{g})$ are 0-conformally equivalent.

To describe generalized conformal structures, let us introduce contrast functions. Let ρ be a function on $M \times M$. We define a function on M by

$$\rho[X_1 \cdots X_i | Y_1 \cdots Y_j](p) = (X_1)_p \cdots (X_i)_p (Y_1)_q \cdots (Y_j)_q \rho(p\|q)|_{p=q},$$

where $X_1, \cdots X_i, Y_1 \cdots Y_j$ are arbitrary vector fields on M. We call ρ a *contrast function* on M if

$$\rho(p\|p) = 0 \quad (p \in M),$$
$$\rho[X|] = \rho[|X] = 0,$$
$$h(X, Y) := -\rho[X|Y] \quad \text{is a semi-Riemannian metric on } M.$$

We remark that the canonical divergence on a dually flat space is a typical example of contrast function.

For a given contrast function ρ on M, we can define a torsion-free affine connection by the following formula:

$$h(\nabla_X Y, Z) := -\rho[XY|Z].$$

The triplet (M, ∇, h) is a statistical manifold. We say that (M, ∇, h) is *induced* from the contrast function ρ. If we exchange the arguments as $\rho^*(p\|q) := \rho(q\|p)$, then ρ^* is also a contrast function and induces the dual statistical manifold (M, ∇^*, h). For geometry of contrast functions, the following results are known ([7, 8]).

Proposition 1.4. *Let ρ and $\bar{\rho}$ be contrast functions on M, and let λ be a function on M. Suppose that (M, ∇, h) and $(M, \bar{\nabla}, \bar{h})$ are statistical manifolds induced from ρ and $\bar{\rho}$, respectively.*

(1) *If $\bar{\rho}(p\|q) = e^{\lambda(p)} \rho(p\|q)$, then two statistical manifolds (M, ∇, h) and $(M, \bar{\nabla}, \bar{h})$ are (-1)-conformally equivalent.*

(2) *If $\bar{\rho}(p\|q) = e^{\lambda(q)} \rho(p\|q)$, then two statistical manifolds (M, ∇, h) and $(M, \bar{\nabla}, \bar{h})$ are 1-conformally equivalent.*

2. Geometry for q-exponential families

In this section, we discuss geometry of q-exponential families. A q-exponential family is a generalization of the standard exponential family. We will consider conformal relations between the standard information geometry and the q-Fisher geometry.

2.1. *The q-escort probability and the q-expectation*

To begin with, we review the notion of the escort probability and the q-expectation. Suppose that $p(x)$ is a probability distribution on $\mathcal{X}$. For a fixed number q, we define the q-*escort distribution* $P_q(x)$ of $p(x)$ by

$$P_q(x) := \frac{1}{\Omega_q(p)} p(x)^q, \quad \Omega_q(p) := \int_{\mathcal{X}} p(x)^q dx.$$

Let $f(x)$ be a random variable on $\mathcal{X}$. The q-*expectation* of $f(x)$ is the expectation with respect to the q-escort distribution, that is,

$$E_{q,p}[f(x)] := \int_{\mathcal{X}} f(x) P_q(x) dx = \frac{1}{\Omega_q(p)} \int_{\mathcal{X}} f(x) p(x)^q dx.$$

If the sample space $\mathcal{X}$ is discrete, the q-escort distribution or the q-expectation can be defined by replacing the integral $\int \cdots dx$ with the sum $\sum_{x \in \mathcal{X}}$.

2.2. *The q-exponential family*

Next, we define the q-exponential and the q-logarithm. Suppose that q is a fixed positive number. Then the q-*exponential function* is defined by

$$\exp_q x := \begin{cases} (1 + (1-q)x)^{\frac{1}{1-q}}, q \neq 1, & (1 + (1-q)x > 0), \\ \exp x, & q = 1, \end{cases} \tag{6}$$

and the q-*logarithm function* by

$$\log_q x := \begin{cases} \frac{x^{1-q}-1}{1-q}, q \neq 1, & (x > 0), \\ \log x, & q = 1. \end{cases}$$

If we consider the limit $q \to 1$, the q-exponential and the q-logarithm recover the standard exponential and the standard logarithm, respectively. For simplicity, we assume that the variable x in (6) satisfy the condition $1 + (1-q)x > 0$ if we consider q-exponential function. Hence q-exponential and q-logarithm function are always mutually inverse functions.

Definition 2.1. A statistical model $S_q = \{p(x,\theta) \mid \theta \in \Theta \subset \boldsymbol{R}^n\}$ is called a *q-exponential family* if

$$S_q := \left\{ p(x,\theta) \ \middle| \ p(x;\theta) = \exp_q \left[\sum_{i=1}^{n} \theta^i F_i(x) - \psi(\theta) \right] \right\},$$

where $F_1(x), \ldots, F_n(x)$ are random variables on the sample space $\mathcal{X}$, and $\psi(\theta)$ is a function on the parameter space Θ.

The information geometric structure of the q-exponential family is closely related to the $(1 - 2q)$- and the $(2q - 1)$-connections. Hence we fix the relations of two parameters q and α as $1 - 2q = \alpha$.

Example 2.1 (q-normal distributions). *A q-normal distribution is the probability distribution defined by the following formula:*

$$p(x; \mu, \sigma) = \frac{1}{Z_{q,\sigma}} \left[1 - \frac{1-q}{3-q} \frac{(x-\mu)^2}{\sigma^2} \right]_+^{\frac{1}{1-q}},$$

where $[*]_+ = \max\{0, *\}$, $\{\mu, \sigma\}$ *are parameters* $-\infty < \mu < \infty, 0 < \sigma < \infty$, *and* $Z_{q,\sigma}$ *is the normalization defined by*

$$Z_{q,\sigma} = \begin{cases} \frac{\sqrt{3-q}}{\sqrt{1-q}} \, \mathrm{Beta}\left(\frac{2-q}{1-q}, \frac{1}{2}\right) \sigma, & (-\infty < q < 1), \\ \frac{\sqrt{3-q}}{\sqrt{q-1}} \, \mathrm{Beta}\left(\frac{3-q}{2(q-1)}, \frac{1}{2}\right) \sigma, & (1 \le q < 3). \end{cases}$$

Set

$$\theta^1 = \frac{2}{3-q} Z_{q,\sigma}^{q-1} \cdot \frac{\mu}{\sigma^2},$$

$$\theta^2 = -\frac{1}{3-q} Z_{q,\sigma}^{q-1} \cdot \frac{1}{\sigma^2},$$

$$\psi(\theta) = -\frac{(\theta^1)^2}{4\theta^2} - \frac{Z_{q,\sigma}^{q-1} - 1}{1-q},$$

then

$$\log_q p_q(x) = \frac{1}{1-q}(p^{1-q} - 1)$$

$$= \frac{1}{1-q} \left\{ \frac{1}{Z_{q,\sigma}^{1-q}} \left(1 - \frac{1-q}{3-q} \frac{(x-\mu)^2}{\sigma^2}\right) - 1 \right\}$$

$$= \frac{2\mu Z_{q,\sigma}^{q-1}}{(3-q)\sigma^2} x - \frac{Z_{q,\sigma}^{q-1}}{(3-q)\sigma^2} x^2 - \frac{Z_{q,\sigma}^{q-1}}{3-q} \cdot \frac{\mu^2}{\sigma^2} + \frac{Z_{q,\sigma}^{q-1} - 1}{1-q}$$

$$= \theta^1 x + \theta^2 x^2 - \psi(\theta).$$

This implies that the set of q-normal distributions is a q-exponential family.

We remark that q-normal distributions include several important probability distributions. If $q = 1$, then the q-normal distribution is the normal distribution, of course. If $q = 2$, then the distribution is the Cauchy distribution. If $q = 1 + 1/(n+1)$, then the distribution is Student's t-distribution. We also remark that mathematical properties of q-normal distributions have been obtained by several authors. See [16, 17], for example.

Example 2.2 (discrete distributions). *Suppose that the sample space $\mathcal{X}$ is a finite discrete set. Then the set of all probability distributions on $\mathcal{X}$ is given by*

$$S_n = \left\{ p(x, \eta) \;\middle|\; \eta_i > 0, \sum_{i=1}^{n+1} \eta_i = 1, \; p(x; \eta) = \sum_{i=1}^{n+1} \eta_i \delta_i(x) \right\},$$

where $\delta_i(x)$ equals one if $x = i$ and zero otherwise. Set

$$\theta^i = \frac{1}{1-q} \left\{ (\eta_i)^{1-q} - (\eta_{n+1})^{1-q} \right\},$$

$$\psi(\theta) = -\log_q \eta_{n+1},$$

then we obtain

$$\log_q p_q(x) = \frac{1}{1-q} \left\{ p^{1-q}(x) - 1 \right\}$$

$$= \frac{1}{1-q} \left\{ \sum_{i=1}^{n+1} (\eta_i)^{1-q} \delta_i(x) - 1 \right\}$$

$$= \frac{1}{1-q} \left\{ \sum_{i=1}^{n} \left((\eta_i)^{1-q} - (\eta_{n+1})^{1-q} \right) \delta_i(x) + (\eta_{n+1})^{1-q} - 1 \right\}$$

$$= \sum_{i=1}^{n} \theta^i \delta_i(x) - \psi(\theta).$$

This implies that the set of discrete distributions is a q-exponential family. We note that this also holds in the case $q = 1$, that is, the set of discrete distribution is an exponential family.

2.3. *Geometry for q-exponential families*

For a q-exponential family $S_q = \{p(x; \theta)\}$, we assume that the potential function ψ is strictly convex. We define the *q-Fisher metric* and the *q-cubic form* in the same manner as exponential families (1) and (2):

$$g_{ij}^q(\theta) = \partial_i \partial_j \psi(\theta),$$

$$C_{ijk}^q(\theta) = \partial_i \partial_j \partial_k \psi(\theta).$$

Since g^q is a Hessian metric on $\{S_q\}$, we can define a flat affine connection $\nabla^{q(e)} = \nabla^{q(1)}$ by

$$g^q(\nabla_X^{q(e)}Y, Z) = g^q(\nabla_X^{q(0)}Y, Z) - \frac{1}{2}C^q(X, Y, Z),$$

where $\nabla^{q(0)}$ is the Levi-Civita connection with respect to the q-Fisher metric g^q. In this case, the parameters $\{\theta^i\}$ is a $\nabla^{q(e)}$-affine coordinate system. We denote by $\nabla^{q(m)}$ the dual connection of $\nabla^{q(e)}$ with respect to g^q. We call $\nabla^{q(e)}$ the *q-exponential connection* and $\nabla^{q(m)}$ the *q-mixture connection*.

Since $\nabla^{q(e)}$ is flat, then $\nabla^{q(m)}$ is also flat. Hence we immediately obtain the following proposition.

Proposition 2.1. *Let S_q be a q-exponential family. Then the tetrad $(S_q, g^q, \nabla^{q(e)}, \nabla^{q(m)})$ is a dually flat space.*

Let S_q be a q-exponential family. From a direct calculation, we have

$$\partial_i p(x; \theta) = p(x; \theta)^q (F_i(x) - \partial_i \psi(\theta)),$$

where $\partial_i = \partial/\partial\theta^i$. Since $\int_\mathcal{X} \partial_i p(x, \theta)dx = \partial_i \int_\mathcal{X} p(x, \theta)dx = 0$, we obtain

$$\partial_i \psi(\theta) = \frac{1}{\Omega_q(p)} \int_\mathcal{X} F_i(x)p(x;\theta)^q dx = \int_\mathcal{X} F_i(x)P_q(x)dx.$$

This implies that the q-mixture parameters are given by the q-expectation of the random variables $\{F_i\}$. Hence we conclude

Proposition 2.2. *Let S_q be a q-exponential family. Then the q-mixture parameters $\{\eta_i\}$ are given by the q-expectation of the random variables $F_i(x)$, that is,*

$$\eta_i = \frac{\partial}{\partial\theta^i}\psi(\theta) = \int_\mathcal{X} F_i(x)P_q(x;\theta)dx.$$

Next, we consider relations between the standard Fisher structure and the q-Fisher structure from the viewpoint of contrast functions.

For a q-exponential distribution S_q, we denote by ρ_q the canonical divergence (5).

Proposition 2.3. *Let S_q be a q-exponential family. Then the canonical divergence ρ_q on S_q is given by*

$$\rho_q(p(\theta')\|p(\theta)) = E_{q,p(\theta)}[\log_q p(\theta) - \log_q p(\theta')].$$

Proof. Since $(S_q, g^q, \nabla^{q(e)}, \nabla^{q(m)})$ is a dually flat space, the q-Fisher metric has a potential function ψ. We denote ϕ by the dual potential function of ψ. For probability distributions $p(\theta)$ and $p(\theta')$ in S_q, using the Legendre duality (3), we obtain

$$E_{q,p(\theta)}[\log_q p(\theta) - \log_q p(\theta')]$$

$$= \int_{\mathcal{X}} \left(\sum_{i=1}^{n} \theta^i F_i(x) - \psi(\theta) - \sum_{i=1}^{n} (\theta')^i F_i(x) + \psi(\theta') \right) P_q(x; \theta) dx$$

$$= \sum_{i=1}^{n} \theta^i \eta_i - \psi(\theta) - \sum_{i=1}^{n} (\theta')^i \eta_i(x) + \psi(\theta')$$

$$= \psi(\theta') + \phi(\theta) - \sum_{i=1}^{n} (\theta')^i \eta_i$$

$$= \rho_q(p(\theta')\|p(\theta)). \qquad \square$$

We remark that the canonical divergence $\rho_q(p(\theta)\|p(\theta'))$ induces the statistical manifold $(S_q, \nabla^{q(e)}, g^q)$ and the dual divergence $\rho_q^*(p(\theta)\|p(\theta')) := \rho_q(p(\theta')\|p(\theta))$ induces $(S_q, \nabla^{q(m)}, g^q)$. The q-exponential family also has another divergence, called the *divergence of Csiszár type* ρ_q^C, which is defined by

$$\rho_q^C(p(\theta)\|p(\theta')) := \frac{1}{1-q} \left\{ 1 - \int_{\mathcal{X}} p(\theta)^q p(\theta')^{1-q} dx \right\}.$$

This is essentially equivalent to the q times of the $(1-2q)$-divergence in information geometry. The divergence $(1/q)\rho_q^C$ induces the statistical manifold $(S_q, \nabla^{(1-2q)}, g^F)$.

Proposition 2.4. *Suppose that ρ_q and ρ_q^C are the canonical divergence and the divergence of Csiszár type on a q-exponential family, respectively. Denote by $\Omega_q(p(\theta))$ the normalization for the q-escort distribution of $p(\theta)$. Then ρ_q and ρ_q^C satisfy*

$$\rho_q(p(\theta')\|p(\theta)) = \frac{1}{\Omega_q(p(\theta))} \rho_q^C(p(\theta)\|p(\theta')).$$

Proof. From Proposition 2.3 we obtain

$$\rho_q(p(\theta')\|p(\theta)) = E_{q,p(\theta)}[\log_q p(\theta) - \log_q p(\theta')]$$

$$= \int_{\mathcal{X}} \left(\frac{p(\theta)^{1-q} - 1}{1-q} - \frac{p(\theta')^{1-q} - 1}{1-q} \right) \frac{p(\theta)^q}{\Omega_q(p(\theta))} dx$$

$$= \frac{1 - \int_{\mathcal{X}} p(\theta)^q p(\theta')^{1-q} dx}{(1-q)\Omega_q(p(\theta))}$$

$$= \frac{1}{\Omega_q(p(\theta))} \rho_q^C(p(\theta)\|p(\theta')). \qquad \square$$

Theorem 2.1. *For a q-exponential family $\{S_q\}$, statistical manifolds $(S_q, \nabla^{q(e)}, g^q)$ and $(S_q, \nabla^{(2q-1)}, g^F)$ are 1-conformally equivalent.*

Proof. Recall that $\rho_q(p(\theta)\|p(\theta'))$ induces $(S_q, \nabla^{q(e)}, g^q)$. From duality of contrast function, $(1/q)\rho_q^{C*}(p(\theta)\|p(\theta')) = (1/q)\rho_q^C(p(\theta')\|p(\theta))$ induces $(S_q, \nabla^{(2q-1)}, g^F)$. From Proposition 2.4, we have

$$\rho_q(p(\theta)\|p(\theta')) = \frac{1}{\Omega_q(p(\theta'))} \rho_q^C(p(\theta')\|p(\theta)) = \frac{1}{\Omega_q(p(\theta'))} \rho_q^{C*}(p(\theta)\|p(\theta')).$$

This implies that two statistical manifolds are 1-conformally equivalent from Proposition 1.4. $\square$

We remark that this theorem was already obtained in the case that the sample space $\mathcal{X}$ is discrete ([13, 14]). For the dual statistical manifolds, we obtain the following corollary immediately.

Corollary 2.1. *For a q-exponential family $\{S_q\}$, two statistical manifolds $(S_q, \nabla^{q(m)}, g^q)$ and $(S_q, \nabla^{(1-2q)}, g^F)$ are (-1)-conformally equivalent.*

Since $(S_q, g^q, \nabla^{q(e)}, \nabla^{q(m)})$ is dually flat, we also obtain the following corollary.

Corollary 2.2. *For a q-exponential family $\{S_q\}$, the statistical manifold $(S_q, \nabla^{(2q-1)}, g^F)$ is 1-conformally flat, and $(S_q, \nabla^{(1-2q)}, g^F)$ is (-1)-conformally flat.*

For generalization of exponential families, several results have been obtained in more generalized frameworks (see [4, 10, 11, 12]). If we consider relations between the standard Fisher geometry and dually flat structures for them as in our paper, some suitable assumptions may be required.

3. An application to statistical inferences

In this section, we discuss an application of geometry of q-exponential families to statistical inferences along the author's explanatory report [9].

3.1. *Generalization of independence*

At first, let us recall the independence of random variables. Suppose that X and Y are random variables which belong to probability density functions $p_1(x)$ and $p_2(y)$, respectively. We say that X and Y are *independent* if the joint probability density function $p(x, y)$ is defined by the product of the marginal probability density functions, that is,

$$p(x, y) = p_1(x)p_2(y).$$

We assume that $p_1(x)$ and $p_2(y)$ are positive everywhere on the sample space. Then the above equation can be written as follows:

$$p(x, y) = p_1(x)p_2(y) = \exp\left[\log p_1(x) + \log p_2(x)\right].$$

This implies that the notion of independence depends on the duality of the exponential function and the logarithm function, or the law of exponents. Hence we can generalize the notion of independence from the viewpoint of q-exponential functions.

For a fixed positive number q, we assume that $x > 0, y > 0$ and $x^{1-q} + y^{1-q} - 1 > 0$. The *q-product* [2] of x and y is defined by

$$x \otimes_q y := \left[x^{1-q} + y^{1-q} - 1\right]^{\frac{1}{1-q}}.$$

The following properties follow from the definition of q-product.

$$\exp_q x \otimes_q \exp_q y = \exp_q(x + y),$$
$$\log_q(x \otimes_q y) = \log_q x + \log_q y.$$

Let us define the notion of q-independence. We say that X and Y are *q-independent with m-normalization* (mixture normalization) if the joint probability density function $p_q(x, y)$ is defined by the q-product of the marginal probability density functions, that is,

$$p_q(x, y) = \frac{p_1(x) \otimes_q p_2(y)}{Z_{p_1, p_2}},$$

where Z_{p_1, p_2} is the normalization defined by

$$Z_{p_1, p_2} = \int\int_{\mathcal{X}\mathcal{Y}} p_1(x) \otimes_q p_2(y) dx dy.$$

Since the q-product of probability density functions $p_1(x) \otimes_q p_2(y)$ is not a probability density in general, a suitable normalization is required [4].

3.2. Geometry for q-likelihood estimators

Let $S = \{p(x;\xi)|\xi \in \Xi\}$ be a statistical model, and let $\{x_1, \ldots, x_N\}$ be N-independent observations generated from a probability density function $p(x;\xi) \in S$. We define the *q-likelihood function* [16] $L_q(\xi)$ by

$$L_q(\xi) = p(x_1;\xi) \otimes_q p(x_2;\xi) \otimes_q \cdots \otimes_q p(x_N;\xi).$$

In the case $q \to 1$, the q-likelihood function L_q is the standard likelihood function on Ξ. Though L_q may not be a probability density on Ξ, we regard L_q as a generalization of the likelihood function.

Since q-logarithm functions are strictly increasing, it is equivalent to consider the q-logarithm q-likelihood function [3]

$$\log_q L_q(\xi) = \sum_{i=1}^{N} \log_q p(x_i;\xi).$$

We say that $\hat{\xi}$ is the *maximum q-likelihood estimator* if

$$\hat{\xi} = \arg \max_{\xi \in \Xi} L_q(\xi) \quad \left(= \arg \max_{\xi \in \Xi} \log_q L_q(\xi) \right).$$

Now let us consider q-likelihood estimator for q-exponential families. Let S_q be a q-exponential family and let M be a curved q-exponential family in S. Suppose that $\{x_1, \ldots, x_N\}$ are N-independent observations generated from $p(x;u) = p(x;\theta(u)) \in M$.

Then the q-likelihood function is calculated as

$$\log_q L_q(u) = \sum_{j=1}^{N} \log_q p(x_j;u) = \sum_{j=1}^{N} \left\{ \sum_{i=1}^{n} \theta^i(u) F_i(x_j) - \psi(\theta(u)) \right\}$$

$$= \sum_{i=1}^{n} \theta^i(u) \sum_{j=1}^{N} F_i(x_j) - N\psi(\theta(u)).$$

The q-logarithm q-likelihood equation is

$$\partial_i \log_q L_q(u) = \sum_{j=1}^{N} F_i(x_j) - N\partial_i \psi(\theta(u)) = 0.$$

Thus, the q-likelihood estimator for S is given by

$$\hat{\eta}_i = \frac{1}{N} \sum_{j=1}^{N} F_i(x_j).$$

On the other hand, the canonical divergence can be calculated as

$$\rho_q^*(p(\hat{\eta})\|p(\theta(u))) = \rho_q(p(\theta(u))\|p(\hat{\eta}))$$
$$= \psi(\theta(u)) + \phi(\hat{\eta}) - \sum_{i=1}^{n} \theta^i(u)\hat{\eta}_i$$
$$= \phi(\hat{\eta}) - \frac{1}{N} \log_q L_q(u).$$

Hence the q-likelihood is maximum if and only if the canonical divergence is minimum. In the same arguments as the standard exponential families, we can say that the q-likelihood estimator is the orthogonal projection from $\hat{\eta}$ to the model distribution M with respect to $\nabla^{q(m)}$-geodesic. Hence the q-likelihood estimator is a quite natural generalization of the likelihood estimator from the viewpoint of differential geometry.

We remark that the q-likelihood can be generalized by U-geometry. The notion of independence is related to geometric structures on the sample space [4].

Acknowledgment

The authors wish to express their sincere gratitude to the referee for his carefully reading and for his apropos comments of the paper.

The first named author is partially supported by The Toyota Physical and Chemical Research Institute and by Grant-in-Aid for Encouragement of Young Scientists (B) No. 19740033, Japan Society for the Promotion of Science.

References

1. S. Amari and H. Nagaoka, *Methods of information geometry*, Amer. Math. Soc., Providence, Oxford University Press, Oxford, 2000.
2. E.P. Borgesa, *A possible deformed algebra and calculus inspired in nonextensive thermostatistics*, Phys. A, **340**(2004), 95–101.
3. D. Ferrari and Y. Yang, *Maximum Lq-likelihood estimation*, Ann. Statist. **38**(2010), 753–783.
4. Y. Fujimoto and N. Murata, *A Generalization of Independence in Naive Bayes Model*, Lecture Notes in Computer Science, **6283**(2010), 153–161.
5. T. Kurose, *Conformal-projective geometry of statistical manifolds*, Interdiscip. Inform. Sci., **8**(2002), 89–100.
6. H. Matsuzoe, *On realization of conformally-projectively flat statistical manifolds and the divergences*, Hokkaido Math. J., **27**(1998), 409–421
7. H. Matsuzoe, *Geometry of contrast functions and conformal geometry*, Hiroshima Math. J., **29**(1999), 175–191.

8. H. Matsuzoe, *Computational Geometry from the Viewpoint of Affine Differential Geometry*, Lecture Notes in Computer Science **5416**(2009), 103–123.
9. H. Matsuzoe, *Geometry for statistical inferences in complex systems*, Toyota Research Report, **63**(2011), 177–180.
10. J. Naudts, *Estimators, escort probabilities, and φ-exponential families in statistical physics*, JIPAM. J. Inequal. Pure Appl. Math., **5**(2004), Article 102 (electronic).
11. J. Naudts, *Generalised exponential families and associated entropy functions*, Entropy, **10**(2008), 131–149.
12. J. Naudts, *Generalised Thermostatistics*, Springer, 2011.
13. A. Ohara, H. Matsuzoe and S. Amari, *A dually flat structure on the space of escort distributions*, J. Phys.: Conf. Ser. **201**(2010), No. 012012 (electronic).
14. A. Ohara, H. Matsuzoe and S. Amari, *Dually flat structure with escort probability and its application to alpha-Voronoi diagrams*, preprint, arXiv:1010.4965 [stat-mech].
15. H. Shima, *The Geometry of Hessian Structures*, World Scientific, 2007.
16. H. Suyari and M. Tsukada, *Law of Error in Tsallis Statistics*, IEEE Trans. Inform. Theory, **51**(2005), 753–757.
17. M. Tanaka, *Meaning of an escort distribution and τ-transformation*, J. Phys.: Conf. Ser. **201**(2010), No 012007 (electronic).
18. C. Tsallis, *Introduction to Nonextensive Statistical Mechanics: Approaching a Complex World*, Springer, New York, 2009.

Received January 31, 2011
Revised April 16, 2011

Proceedings of the 2nd International
Colloquium on Differential Geometry
and its Related Fields
Veliko Tarnovo, September 6–10, 2010

73

SASAKIAN MAGNETIC FIELDS
ON HOMOGENEOUS REAL HYPERSURFACES
IN A COMPLEX HYPERBOLIC SPACE

Tuya BAO*

*Division of Mathematics and Mathematical Science, Nagoya Institute of Technology,
Nagoya 466-8555, Japan
College of Mathematics, Inner Mongolia University for the Nationalities,
Tongliao, Inner Mongolia, 028043, People's Republic of China
E-mail: btyngy@yahoo.co.jp*

On a real hypersurface in a Kähler manifold we have a natural closed 2-form
associated with the induced almost contact metric structure. Constant multiples of this form are called Sasakian magnetic fields. In this article we explain
the behavior of trajectories for these magnetic fields which are also circles or
curves of order 2 on homogeneous Hopf real hypersurfaces of constant principal
curvatures in a complex hyperbolic space.

Keywords: Sasakian magnetic fields, Trajectories, Circles, Curves of order 2,
Homogeneous real hypersurfaces, Structure torsion, Length spectrum.

1. Introduction

In this paper, we give an overview of the author's recent joint works on
real hypersurfaces in a complex hyperbolic space from the viewpoint of
curve-theory. In order to get shapes of Riemannian manifolds it is important to investigate their curves. It is needless to say that the shape of
a Riemannian manifold give a great influence on properties of curves on
this manifold. Conversely, properties of some family of curves show some
properties of the base manifold. One of natural ways in this direction is
to study Riemannian manifolds by investigating properties of geodesics.
We are hence interested in studying Riemannian manifolds with additional
geometric structures from such a point of view.

On a real hypersurface in a Kähler manifold we have an induced almost
contact metric structure $(\phi, \xi, \eta, \langle\ ,\ \rangle)$ (see §2). Associated with this structure we consider a family of trajectories for Sasakian magnetic fields, which

*The author is partially supported by NGK Foundation for International Students.

are smooth curves whose accelerations are parallel to ϕ-images of velocity vectors. We consider whether they are "elemental" on a real hypersurface from curve theoretic point of view through the Frenet-Serret formula. We shall consider geodesics as most elemental curves because they do not have accelerations. Since the family of all trajectories for Sasakian magnetic fields on real hypersurfaces is too complicated, we take circles, and more generally, curves of order 2 (see §4 for definitions) as curves which are elemental next to geodesics. Our consideration gives model spaces in the study of real hypersurfaces in Kähler manifolds by investigating trajectories for Sasakian magnetic fields.

The author is grateful to Professor T. Adachi for his advice in preparing this article.

2. Kähler and Sasakian magnetic fields

As a generalization of static magnetic field, we say a closed 2-form $\mathbb{B}$ on a Riemannian manifold M to be a *magnetic field* (see [18] for example). We define a skew-symmetric operator $\Omega_{\mathbb{B}} : TM \to TM$ by $\langle v, \Omega_{\mathbb{B}}(w) \rangle = \mathbb{B}(v, w)$ for all $v, w \in T_p M$ at an arbitrary point $p \in M$. We say a smooth curve γ parameterized by its arclength to be a *trajectory* if it satisfies the ordinary differential equation $\nabla_{\dot\gamma}\dot\gamma = \Omega_{\mathbb{B}}(\dot\gamma)$. Here, $\nabla_{\dot\gamma}$ denotes the covariant differentiation along γ with respect to the Riemannian connection ∇ on M. A trajectory shows a motion of a charged particle of unit mass under this magnetic field. For the trivial magnetic field, which is the null 2-form, its trajectories are geodesics, because the associated skew-symmetric operator Ω is trivial. Just like geodesics induce the geodesic flow, trajectories induce a dynamical system on the unit tangent bundle. We are hence interested in the relationship between properties of trajectories and that of a base manifold.

When we treat magnetic fields, it might be easy to treat if strengths of magnetic fields do not depend on positions and directions. We call a magnetic field $\mathbb{B}$ uniform if its skew-symmetric operator satisfies $\nabla \Omega_{\mathbb{B}} \equiv 0$. A typical example of uniform magnetic fields is a constant multiple of the Kähler form on a Kähler manifold. It is called a Kähler magnetic field. For a Kähler magnetic field $\mathbb{B}_\kappa = \kappa \mathbb{B}_J$ on a Kähler manifold $\widetilde{M}$ with complex structure J and Kähler form $\mathbb{B}_J$, its equation for trajectories is of the form $\nabla_{\dot\gamma}\dot\gamma = \kappa J\dot\gamma$. In his papers ([1] and its sequels), Adachi studied properties of trajectories for Kähler magnetic fields by comparing those of geodesics.

In the study of Kähler magnetic fields, complex space forms, which are complex projective spaces, complex Euclidean spaces and complex hyperbolic spaces, play as model spaces. This corresponds to the fact that model spaces for the study of geodesics are real space forms, which are standard spheres, Euclidean spaces and real hyperbolic spaces. For Kähler magnetic fields some comparison theorems were obtained (see [5] and its references). Since Kähler manifolds are of real even dimensional, the author is interested in a study of magnetic fields on odd dimensional manifolds. As odd dimensional objects corresponding to complex space forms we have Sasakian space forms. Geodesic spheres, horospheres and tubes around totally geodesic complex hypersurfaces in complex space forms are Sasakian space forms if their radii and holomorphic sectional curvatures of ambient spaces satisfy some relation. Thus it is natural to come to study real hypersurfaces in Kähler manifolds.

A real hypersurface M in a Kähler manifold $\widetilde{M}$ with complex structure J and Riemannian metric $\langle\ ,\ \rangle$ admits an almost contact metric structure. It is a quartet $(\phi, \xi, \eta, \langle\ ,\ \rangle)$ of a vector field ξ, a (1,1)-tensor ϕ, a function η and an induced metric on M. They are defined as

$$\xi = -J\mathcal{N}, \quad \eta(v) = \langle v, \xi \rangle, \quad \phi(v) = Jv - \eta(v)\mathcal{N}$$

for each $v \in TM$ with a unit normal vector field $\mathcal{N}$ of M in $\widetilde{M}$. We call ξ and ϕ the characteristic vector field and the characteristic tensor, respectively. We define a 2-form $\mathbb{F}_\phi$ on M by $\mathbb{F}_\phi(v, w) = \langle v, \phi(w) \rangle$. One can easily check that it is a closed 2-form, hence is a magnetic field (see [8]). We say a constant multiple $\mathbb{F}_\kappa = \kappa\mathbb{F}_\phi$ to be a *Sasakian magnetic field*. The equation of trajectories for $\mathbb{F}_\kappa$ is hence of the form $\nabla_{\dot\gamma}\dot\gamma = \kappa\phi\dot\gamma$. Unfortunately, Sasakian magnetic fields are not uniform because we have the following equalities:

$$\nabla_X \xi = \phi AX, \quad (\nabla_X \phi)Y = \langle Y, \xi \rangle AX - \langle AX, Y \rangle \xi, \tag{1}$$

for arbitrary vector fields X, Y on M. Here A denotes the shape operator of M in $\widetilde{M}$. Therefore, even Kähler magnetic fields and Sasakian magnetic fields are quite resemble in definitions, there are many difference in their properties.

3. Real hypersurfaces in a complex hyperbolic space

In this paper we restrict ourselves to real hypersurfaces in a complex hyperbolic space $\mathbb{C}H^n(-c)$ of constant holomorphic sectional curvature $-c$. We say a real hypersurface in a Kähler manifold to be a Hopf hypersurface if its characteristic vector field ξ is principal at each point. Here, we recall that eigenvalues and eigenvectors of the shape operator of M are called principal curvatures and principal vectors, respectively. In $\mathbb{C}H^n(-c)$, it is known that a Hopf hypersurface all of whose principal curvatures are constant on this surface is locally congruent to one of the following (see [12]):

(A$_0$) a horosphere HS,
(A$_1$) a geodesic sphere $G(r)$ of radius r,
(A$_1$) a tube $T(r)$ of radius r around totally geodesic $\mathbb{C}H^{n-1}(-c)$,
(A$_2$) a tube $T_\ell(r)$ of radius r around totally geodesic $\mathbb{C}H^\ell(-c)$, where $1 \le \ell \le n-2$,
(B) a tube $R(r)$ of radius r around totally geodesic $\mathbb{R}H^n(-c/4)$.

These real hypersurfaces are called hypersurfaces of type (A$_0$), (A$_1$), (A$_1$), (A$_2$) and (B), respectively. We note that both geodesic spheres and tubes around totally geodesic $\mathbb{C}H^{n-1}$ are called of type (A$_1$). Summarizing first four kinds of real hypersurfaces up, we call them hypersurfaces of type (A). We give principal curvatures of these real hypersurfaces in Table 1 (see [12, 17]). For a real hypersurface M of type either (A) or (B), we denote by ν_M the principal curvature of ξ. Hence λ_M and μ_M denote principal curvatures for principal vectors orthogonal to ξ. We denote by V_λ, V_μ the subbundles of all principal curvature vectors which are orthogonal to ξ and are associated with $\lambda = \lambda_M, \mu = \mu_M$, respectively. For hypersurfaces of type (A), they satisfy $\phi(V_\lambda) = V_\lambda$ and $\phi(V_\mu) = V_\mu$, hence their shape operators and characteristic tensors are simultaneously diagonalizable (i.e. $A\phi = \phi A$). For hypersurfaces of type (B), they satisfy $\phi(V_\lambda) = V_\mu$ and $\phi(V_\mu) = V_\lambda$.

Table 1. Principal curvatures of homogeneous real hypersurfaces of types (A) and (B) in $\mathbb{C}H^n(-c)$.

M	λ_M	μ_M	ν_M
HS	$\sqrt{c}/2$	—	$\sqrt{c}$
$G(r)$	$(\sqrt{c}/2)\coth(\sqrt{c}\,r/2)$	—	$\sqrt{c}\coth(\sqrt{c}\,r)$
$T(r)$	$(\sqrt{c}/2)\tanh(\sqrt{c}\,r/2)$	—	$\sqrt{c}\coth(\sqrt{c}\,r)$
$T_\ell(r)$	$(\sqrt{c}/2)\coth(\sqrt{c}\,r/2)$	$(\sqrt{c}/2)\tanh(\sqrt{c}\,r/2)$	$\sqrt{c}\coth(\sqrt{c}\,r)$
$R(r)$	$(\sqrt{c}/2)\coth(\sqrt{c}\,r/2)$	$(\sqrt{c}/2)\tanh(\sqrt{c}\,r/2)$	$\sqrt{c}\coth(\sqrt{c}\,r)$

4. Circles and curves of order two

In order to explain our results we here recall some terminology on curves. A smooth curve γ on a Riemannian manifold which is parameterized by its arclength is said to be a helix of proper order d (≥ 1) if it satisfies the system of ordinary differential equations

$$\nabla_{\dot\gamma} Y_i = -k_{i-1} Y_{i-1} + k_i Y_{i+1} \qquad (i = 1, \ldots, d)$$

with positive constants $k_1, \ldots, k_{d-1}$ and a field of orthonormal frame $\{Y_1 = \dot\gamma, Y_2, \ldots, Y_d\}$ along γ. Here we set $k_0 = k_d = 0$ and Y_0, Y_{d+1} are null vector fields along γ. We call those constants $k_1, \ldots, k_{d-1}$ and the frame field $\{Y_i\}_{i=1}^d$ *geodesic curvatures* and *Frenet frame* of a helix γ, respectively. We say a smooth curve to be a helix of order d if it is a helix of proper order not greater than d. A helix of order 1 is a geodesic, and a helix of order 2 is called a circle. Thus a circle of positive geodesic curvature k is a smooth curve γ satisfying the differential equations $\nabla_{\dot\gamma}\dot\gamma = kY$ and $\nabla_{\dot\gamma} Y = -k\dot\gamma$ with a unit vector field Y along γ which is orthogonal to $\dot\gamma$. If we rewrite these equations we find that a smooth curve γ parameterized by its arclength is a circle if and only if it satisfies $\nabla_{\dot\gamma}\nabla_{\dot\gamma}\dot\gamma + \|\nabla_{\dot\gamma}\dot\gamma\|^2\dot\gamma = 0$. For the sake of later use we here introduce a bit more generalized notion. We say a smooth curve γ parameterized by its arclength to be a *curve of order* 2 if it satisfies the equation

$$\|\nabla_{\dot\gamma}\dot\gamma\|^2\{\nabla_{\dot\gamma}\nabla_{\dot\gamma}\dot\gamma + \|\nabla_{\dot\gamma}\dot\gamma\|^2\dot\gamma\} = \langle\nabla_{\dot\gamma}\dot\gamma, \nabla_{\dot\gamma}\nabla_{\dot\gamma}\dot\gamma\rangle\nabla_{\dot\gamma}\dot\gamma. \qquad (2)$$

Geodesics clearly satisfy this equation, and so do circles. More generally, if γ is a Frenet curve of order 2, that is a smooth curve parameterized by its arclength which satisfies

$$\nabla_{\dot\gamma}\dot\gamma(t) = k(t)Y(t) \quad \text{and} \quad \nabla_{\dot\gamma} Y(t) = -k(t)\dot\gamma(t) \qquad (3)$$

with some function $k(t)$ and a unit vector field Y along γ orthogonal to $\dot\gamma$, then it also satisfies the equation (2). Therefore we can consider the notion of curves of order 2 is a generalization of the notion of circles. If we say a bit more, when a curve γ of order 2 does not have inflection points, by putting $k(t) = \|\nabla_{\dot\gamma}\dot\gamma(t)\|$ and $Y(y) = (1/k(t))\nabla_{\dot\gamma}\dot\gamma(t)$, we find it satisfies (3).

5. Circular trajectories for Sasakian magnetic fields

We shall now study trajectories for Sasakian magnetic fields. If we consider trajectories for Kähler magnetic fields on a Kähler manifold, as the complex structure J is parallel with respect to the Riemannian connection and satisfies $J^2 = -Id$, we see that they are circles. We are hence interested in whether trajectories for Sasakian magnetic field are circles or not. We first study the first geodesic curvature of trajectories. Let γ be a trajectory for a Sasakian magnetic field $\mathbb{F}_\kappa$ on a real hypersurface M in a Kähler manifold $\widetilde{M}$. Its first geodesic curvature is $|\kappa|\|\phi\dot{\gamma}\| = |\kappa|\sqrt{1 - \langle\dot{\gamma},\xi\rangle^2}$. Thus the quantity $\langle\dot{\gamma},\xi\rangle$ should show some property of γ. We denote this function by ρ_γ and call it the *structure torsion* of γ. It is clear that if this is not a constant function then γ is not a circle. In case the structure torsion of γ is a constant, we calculate the differential of $\phi\dot{\gamma}$ by use of (1):

$$\nabla_{\dot{\gamma}}(\phi\dot{\gamma}) = \rho_\gamma A\dot{\gamma} - \langle A\dot{\gamma},\dot{\gamma}\rangle\xi + \kappa\phi^2\dot{\gamma} = \rho_\gamma A\dot{\gamma} - \langle A\dot{\gamma},\dot{\gamma}\rangle\xi - \kappa(\dot{\gamma} - \rho_\gamma\xi).$$

Hence γ is a circle if it coincides with $-\kappa(1 - \rho_\gamma^2)\dot{\gamma}$ under the assumption that ρ_γ is constant. Thus we obtain the following.

Proposition 5.1. *The feature of a trajectory γ for a non-trivial Sasakian magnetic field $\mathbb{F}_\kappa$ on a real hypersurface in a Kähler manifold is as follows:*

(1) *It is a geodesic if and only if $\rho_\gamma \equiv \pm 1$;*
(2) *It is a circle of positive geodesic curvature if and only if it satisfies both of the following conditions:*
   (C1) *$\rho_\gamma' \equiv 0$ and $\rho_\gamma \neq \pm 1$,*
   (C2) *$-\kappa\rho_\gamma^2\dot{\gamma} + \rho_\gamma A\dot{\gamma} + (\kappa\rho_\gamma - \langle A\dot{\gamma},\dot{\gamma}\rangle)\xi \equiv 0.$*

We here calculate the differential of structure torsion of a trajectory γ by use of (1):

$$\rho_\gamma' = \nabla_{\dot{\gamma}}\langle\dot{\gamma},\xi\rangle = \langle\kappa\phi\dot{\gamma},\xi\rangle + \langle\dot{\gamma},\phi A\dot{\gamma}\rangle = \langle\dot{\gamma},\phi A\dot{\gamma}\rangle = \frac{1}{2}\langle\dot{\gamma},(\phi A - A\phi)\dot{\gamma}\rangle.$$

Thus we find ρ_γ is a constant function if and only if $(A\phi - \phi A)\dot{\gamma}(t)$ is perpendicular to $\dot{\gamma}(t)$ at each t. In particular, on hypersurfaces of type (A) in a complex hyperbolic space, the structure torsion of each trajectory for an arbitrary Sasakian magnetic field is constant, because their shape operators and characteristic tensors are simultaneously diagonalizable (i.e. $A\phi = \phi A$).

We now consider hypersurfaces of types (A_0) and (A_1) in a complex hyperbolic space. We call a trajectory *circular* if it is also a circle of positive geodesic curvature. Since hypersurfaces of types (A_0) and (A_1) have two distinct principal curvatures, Proposition 5.1 shows the following.

Theorem 5.1 ([8]). *Let $\mathbb{F}_\kappa$ be a non-trivial Sasakian magnetic field on a real hypersurface M of type either (A_0) or (A_1) in $\mathbb{C}H^n(-c)$.*

(1) *There are no circular trajectories for $\mathbb{F}_\kappa$ if $0 < |\kappa| \leq \lambda_M$.*
(2) *When $|\kappa| > \lambda_M$, a trajectory γ for $\mathbb{F}_\kappa$ is circular if and only if $\rho_\gamma = \lambda_M/\kappa$. In this case its geodesic curvature is $\sqrt{\kappa^2 + \lambda_M^2}$.*

If we describe them individually, it turns to the followings:

Corollary 5.1. *We consider a non-trivial Sasakian magnetic field $\mathbb{F}_\kappa$ on a horosphere HS in $\mathbb{C}H^n(-c)$.*

(1) *When $0 < |\kappa| \le \sqrt{c}/2$, there are no circular trajectories for $\mathbb{F}_\kappa$.*

(2) *When $|\kappa| > \sqrt{c}/2$, a trajectory γ for $\mathbb{F}_\kappa$ is circular if and only if $\rho_\gamma = \sqrt{c}/(2\kappa)$. In this case its geodesic curvature is $\sqrt{\kappa^2 + (c/4)}$.*

Corollary 5.2. *We consider a non-trivial Sasakian magnetic field $\mathbb{F}_\kappa$ on a geodesic sphere $G(r)$ of radius r in $\mathbb{C}H^n(-c)$.*

(1) *When $0 < |\kappa| \le (\sqrt{c}/2)\coth(\sqrt{c}\,r/2)$, there are no circular trajectories for $\mathbb{F}_\kappa$.*

(2) *When $|\kappa| > (\sqrt{c}/2)\coth(\sqrt{c}\,r/2)$, a trajectory γ for $\mathbb{F}_\kappa$ is circular if and only if $\rho_\gamma = (\sqrt{c}/(2\kappa))\coth(\sqrt{c}\,r/2)$. In this case its geodesic curvature is $\sqrt{\kappa^2 + (c/4)\coth^2(\sqrt{c}\,r/2)}$.*

Corollary 5.3. *We consider a non-trivial Sasakian magnetic field $\mathbb{F}_\kappa$ on a tube $T(r)$ of radius r around $\mathbb{C}H^{n-1}$ in $\mathbb{C}H^n(-c)$.*

(1) *When $0 < |\kappa| \le (\sqrt{c}/2)\tanh(\sqrt{c}\,r/2)$, there are no circular trajectories for $\mathbb{F}_\kappa$.*

(2) *When $|\kappa| > (\sqrt{c}/2)\tanh(\sqrt{c}\,r/2)$, a trajectory γ for $\mathbb{F}_\kappa$ is circular if and only if $\rho_\gamma = (\sqrt{c}/(2\kappa))\tanh(\sqrt{c}\,r/2)$. In this case its geodesic curvature is $\sqrt{\kappa^2 + (c/4)\tanh^2(\sqrt{c}\,r/2)}$.*

By these Proposition 5.1 and Theorem 5.1, we find trajectories for a nontrivial Sasakian magnetic field $\mathbb{F}_\kappa$ on hypersurfaces of types (A_0) and (A_1) are geodesics when $\rho_\gamma = \pm 1$ and are circles of positive geodesic curvature when $\kappa\rho_\gamma = \lambda$. If we consider the other case, as we have $A\dot\gamma = \nu_M\rho_\gamma\xi + \lambda_M(\dot\gamma - \rho_\gamma\xi)$ for the shape operator of a hypersurface M of type either (A_0) or (A_1), the first equality in (1) shows that trajectories on this hypersurface are helices of proper order 3, in general (see Proposition 1 in [8], for more detail).

Next we study trajectories on hypersurfaces of type (A_2) in $\mathbb{C}H^n$. Since these real hypersurfaces have three principal curvatures, we need another invariant to classify trajectories. For a hypersurface M of type (A_2) in $\mathbb{C}H^n$, we take the subbundles V_λ, V_μ given in section 3. These subbundles consist of principal curvature vectors which are orthogonal to ξ and are associated with the principal curvatures $\lambda = \lambda_M, \mu = \mu_M$, respectively. For a trajectory γ on $T_\ell(r)$, we set $\tau_\gamma = \|\mathrm{Proj}_{V_\lambda}(\dot\gamma)\|$ with a projection

$\mathrm{Proj}_{V_\lambda} : TT_\ell(r) = V_\lambda \oplus V_\mu \oplus \mathbb{R}\xi \to V_\lambda$, and call it its *principal torsion*. It is clear that $0 \le \tau_\gamma \le \sqrt{1 - \rho_\gamma^2}$. Since the shape operator A of a hypersurface M in $\mathbb{C}H^n(-c)$ of type (A) satisfies

$$(\nabla_X A)Y = (c/4)\{\langle \phi X, Y \rangle \xi + \langle Y, \xi \rangle \phi X\}$$

for vector fields X, Y on M (see [16] for example), we find $\langle A\dot\gamma, \dot\gamma \rangle$ is constant along a trajectory γ. As we have $\langle A\dot\gamma, \dot\gamma \rangle = \nu_M \rho_\gamma^2 + \lambda_M \tau_\gamma^2 + \mu_M(1 - \rho_\gamma^2 - \tau_\gamma^2)$, we obtain τ_γ is constant along γ. For a trajectory γ for a non-trivial Sasakian magnetic field $\mathbb{F}_\kappa$, the circular condition (C2) turns to

$$\rho_\gamma(\lambda - \kappa\rho_\gamma) \mathrm{Proj}_{V_\lambda}(\dot\gamma) + \rho_\gamma(\mu - \kappa\rho_\gamma) \mathrm{Proj}_{V_\mu}(\dot\gamma)$$
$$+ \{\kappa\rho_\gamma - \kappa\rho_\gamma^3 - \lambda\tau_\gamma^2 - \mu(1 - \rho_\gamma^2 - \tau_\gamma^2)\}\xi = 0,$$

with the projection $\mathrm{Proj}_{V_\mu} : TT_\ell(r) \to V_\mu$. Hence we obtain the following.

Theorem 5.2 ([8]). *We consider a non-trivial Sasakian magnetic field $\mathbb{F}_\kappa$ on a tube $T_\ell(r)$ of radius r around $\mathbb{C}H^\ell(-c)$ $(1 \le \ell \le n - 2)$ in $\mathbb{C}H^n(-c)$.*

(1) *When $0 < |\kappa| \le (\sqrt{c}/2)\tanh(\sqrt{c}r/2)$, there are no circular $\mathbb{F}_\kappa$-trajectories.*

(2) *When $(\sqrt{c}/2)\tanh(\sqrt{c}r/2) < |\kappa| \le (\sqrt{c}/2)\coth(\sqrt{c}r/2)$, a $\mathbb{F}_\kappa$-trajectory γ is circular if and only if it satisfies the condition $\rho_\gamma = (\sqrt{c}/(2\kappa))\tanh(\sqrt{c}r/2)$ and $\tau_\gamma = 0$.*

(3) *When $|\kappa| > (\sqrt{c}/2)\coth(\sqrt{c}r/2)$, a $\mathbb{F}_\kappa$-trajectory γ is circular if and only if it satisfies one of the following:*

   i) *$\rho_\gamma = (\sqrt{c}/2\kappa)\tanh(\sqrt{c}r/2)$ and $\tau_\gamma = 0$,*

   ii) *$\rho_\gamma = (\sqrt{c}/2\kappa)\coth(\sqrt{c}r/2)$ and $\tau_\gamma = \sqrt{1 - \rho_\gamma^2}$.*

In the third, we study trajectories on hypersurfaces of type (B) in $\mathbb{C}H^n$. Since the characteristic tensor of a hypersurface of type (B) maps subbundles of principal curvature vectors as $\phi(V_\lambda) = V_\mu$, structure torsions of trajectories for Sasakian magnetic fields may not be constant functions. We hence consider whether they are curves of order 2 or not. Since we have $\nabla_{\dot\gamma}(\phi\dot\gamma) = \rho_\gamma A\dot\gamma - \langle A\dot\gamma, \dot\gamma \rangle \xi - \kappa(\dot\gamma - \rho_\gamma \xi)$, for a trajectory γ for $\mathbb{F}_\kappa$, substituting this into (2) we find its principal torsion satisfies $(\lambda - \mu)\tau_\gamma^2 = (1 - \rho_\gamma^2)(\kappa\rho_\gamma - \mu)$, which is the same property as for trajectories on hypersurfaces of type (A$_2$). Thus we obtain

Theorem 5.3 ([10]). *When $0 < |\kappa| \le (\sqrt{c}/2)\tanh(\sqrt{c}r/2)$, there are no $\mathbb{F}_\kappa$-trajectories which are also curves of order 2 on a tube $R(r)$ of radius r around $\mathbb{R}H^n(-c/4)$ in $\mathbb{C}H^n(-c)$.*

In order to study more on trajectories on hypersurfaces of type (B), we are interested in behaviors of structure torsions and principal torsions of trajectories. Though there is a study on differentials of shape operators of hypersurfaces of type (B) in [14], it is not easy to apply to our case. We can say that the structure torsion of a trajectory which is also a curve of order 2 is constant if and only if its principal torsion is constant (see [10] for detail).

6. Characterization of hypersurfaces of type (A)

Here we give some characterizations of hypersurfaces of type (A) among real hypersurfaces in $\mathbb{C}H^n$ by the amount of circular trajectories for Sasakian magnetic fields. It is well-known that hypersurfaces of type (A) are characterized by the property that their shape operators and their characteristic tensors are simultaneously diagonalizable (see [17] for example). If a trajectory γ for a Sasakian magnetic field is circular, then Proposition 5.1 shows that ρ_γ is constant, hence γ satisfies $\langle (A\phi - \phi A)\dot\gamma(t), \dot\gamma(t)\rangle \equiv 0$. Since the proof of Proposition 5 in [8] contains an error, we here give corrections.

Theorem 6.1 ([8]). *Let M be a real hypersurface in $\mathbb{C}H^n$. Then the following conditions are mutually equivalent:*

1) *M is congruent to one of HS, $G(r)$ and $T(r)$;*
2) *For every unit tangent vector $v \in UM$ which is neither orthogonal to ξ nor parallel to ξ, there is a circular trajectory for some non-trivial Sasakian magnetic field on M whose initial vector is v.*

Proof. 1) $\Rightarrow$ 2). This is a consequence of Corollaries 5.1, 5.2 and 5.3.

2) $\Rightarrow$ 1). By the second condition we have $\langle (A\phi - \phi A)u, u\rangle = 0$ for arbitrary tangent vector $u \in TM$. Since $A\phi - \phi A$ is symmetric, we find that

$$0 = \langle (A\phi - \phi A)(u+w), u+w\rangle = 2\langle (A\phi - \phi A)u, w\rangle$$

for arbitrary $u, w \in TM$. Hence we get $A\phi - \phi A = 0$ and find that M is of type (A). As hypersurfaces of type (A_2) do not satisfy the second condition, we get the conclusion. $\square$

We shall give other characterization of homogeneous real hypersurfaces in a complex hyperbolic space. First we give a condition a real hypersurface to be Hopf.

Lemma 6.1. *Let M be a real hypersurface in $\mathbb{C}H^n$. Suppose at each point $x \in M$ there is a trajectory for a non-trivial Sasakian magnetic field whose* *initial vector is ξ_x and which is also a geodesic. Then M is a Hopf hypersurface.*

Proof. Let γ_x be a trajectory for $\mathbb{F}_\kappa$ ($\kappa \neq 0$) with $\dot\gamma_x(0) = \xi_x$ which is also a geodesic on M. We then have

$$0 = \nabla_{\dot\gamma_x} \nabla_{\dot\gamma_x} \dot\gamma_x = \kappa \nabla_{\dot\gamma_x}(\phi \dot\gamma_x) = \kappa\{\rho_\gamma A\dot\gamma_x - \langle A\dot\gamma_x, \dot\gamma_x\rangle \xi\}.$$

In particular, at the initial we have $A\xi_x = \langle A\xi_x, \xi_x\rangle \xi_x$ and get the conclusion. $\qquad\square$

In order to characterize real hypersurfaces by properties of trajectories for Sasakian magnetic fields, we may consider in the class of Hopf hypersurfaces. For a real hypersurface M we set $T^0M = \{v \in TM \mid \langle v, \xi\rangle = 0\}$. We denote by $\mathrm{Proj}_0 : TM \to T^0M$ the projection. On a Hopf hypersurface M, the condition (C2) that a trajectory γ for $\mathbb{F}_\kappa$ is circular turns to

$$\begin{cases} \rho_\gamma A\mathrm{Proj}_0(\dot\gamma) = \kappa\rho_\gamma^2 \mathrm{Proj}_0(\dot\gamma), \\ \kappa\rho_\gamma(1 - \rho_\gamma^2) + \rho_\gamma^2\langle A\xi, \xi\rangle - \langle A\dot\gamma, \dot\gamma\rangle = 0. \end{cases} \tag{4}$$

This shows the following results.

Theorem 6.2. *Let M be a Hopf hypersurface in $\mathbb{C}H^n$. Then M is congruent to one of HS, $G(r)$ and $T(r)$ if and only if at each point $x \in M$ there exist unit tangent vectors $v_1, \ldots, v_{2n-2} \in T_x^0M$ satisfying the following properties:*

i) *$v_1, \ldots, v_{2n-2}$ span T_x^0M;*
ii) *For each i ($1 \leq i \leq 2n-2$), there is a circular trajectory γ_i for some Sasakian magnetic field satisfying that $\rho_{\gamma_i} \neq 0$ and $\mathrm{Proj}_0(\dot\gamma_i(0))$ is parallel to v_i;*
iii) *For each j ($2 \leq j \leq 2n-2$), there is a circular trajectory γ_{1j} satisfying that $\rho_{\gamma_{1j}} \neq 0$ and $\mathrm{Proj}_0(\dot\gamma_{1j}(0))$ is parallel to v_1+v_j.*

Proof. The "only if" part is a consequence of Corollaries 5.1, 5.2 and 5.3. We show the "if" part. The first equality in the relation (4) shows that $v_1, \ldots v_{2n-2}$ and $v_1+v_2, \ldots, v_1+v_{2n-2}$ are principal curvature vectors. Since $v_1, \ldots v_{2n-2}$ are linearly independent, we find they have the same principal curvatures. Thus T_x^0M is a vector subspace of principal curvatures. We hence find $A\phi = \phi A$, which lead us to that M is of type (A). As the bundle T_x^0M of a hypersurface of type (A_2) is divided into two principal subbundles, we get the conclusion. $\qquad\square$

Theorem 6.3. *Let M be a Hopf hypersurface in $\mathbb{C}H^n$. Then M is a hypersurface of type* (A) *if and only if at each point $x \in M$ there exist unit tangent vectors $v_1, \ldots, v_{n-1} \in T_x^0 M$ satisfying the following properties:*

i) $v_1, \phi v_1, \ldots, v_{n-1}, \phi v_{n-1}$ *span* $T_x^0 M$;

ii) *For each i $(1 \leq i \leq n-1)$, there are circular trajectories γ_i^+, γ_i^- for some Sasakian magnetic fields $\mathbb{F}_{\kappa_i^+}, \mathbb{F}_{\kappa_i^-}$ satisfying that*

a) $\rho_{\gamma_i^+} \neq 0$, $\rho_{\gamma_i^-} \neq 0$,

b) $\kappa_i^+ \rho_{\gamma_i^+} = \kappa_i^- \rho_{\gamma_i^-}$,

c) $\mathrm{Proj}_0\left(\dot\gamma_i^+(0)\right)$ *is parallel to v_i and $\mathrm{Proj}_0\left(\dot\gamma_i^-(0)\right)$ is parallel to ϕv_i.*

Proof. The "only if" part is a consequence of Corollaries 5.1, 5.2, 5.3 and Theorem 5.2. We show the "if" part. The first equality in the relation (4) shows that v_i and ϕv_i are principal curvature vectors associated with principal curvature $\kappa_i^+ \rho_{\gamma_i^+} = \kappa_i^- \rho_{\gamma_i^-}$. We hence find that each vector subspace of principal curvature vectors in $T_x^0 M$ is invariant under the action of ϕ. Thus we find $A\phi = \phi A$ and M is of type (A). $\qquad\square$

The proof of the above Theorem shows the following.

Theorem 6.4. *Let M be a Hopf hypersurface in $\mathbb{C}H^n$. Then M is congruent to one of HS, $G(r)$ and $T(r)$ if and only if at each point $x \in M$ there exist unit tangent vectors $v_1, \ldots, v_{2n-2} \in T_x^0 M$ and a nonzero constant α satisfying the following properties:*

i) $v_1, \ldots, v_{2n-2}$ *span* $T_x^0 M$;

ii) *For each i $(1 \leq i \leq 2n-2)$, there are circular trajectories γ_i for some Sasakian magnetic fields $\mathbb{F}_{\kappa_i}$ satisfying that $\kappa_i \rho_{\gamma_i} = \alpha$ and $\mathrm{Proj}_0\left(\dot\gamma_i(0)\right)$ is parallel to v_i.*

7. Extrinsic shapes of trajectories

In this section, we study trajectories for Sasakian magnetic fields on hypersurfaces of types (A_0) and (A_1) by studying in the ambient space $\mathbb{C}H^n$. For a curve γ on a real hypersurface M in a Kähler manifold $\widetilde{M}$ we call the curve $\iota \circ \gamma$ with an immersion $\iota : M \to \widetilde{M}$ its *extrinsic shape*.

For a helix σ of proper order d on a Kähler manifold we define its complex torsions τ_{ij} $(1 \leq i < j \leq d)$ by $\tau_{ij} = \langle Y_i, JY_j \rangle$ with its Frenet frame field $\{Y_i\}_{i=1}^d$. When a smooth curve on a manifold M is an orbit of some one parameter family of isometries of M, it is called *Killing*. It is known that a helix on $\mathbb{C}H^n$ is Killing if all its complex torsions are constant functions ([16]). Since trajectories for Kähler magnetic fields on $\mathbb{C}H^n$ have constant complex torsion $\tau_{12} = \pm 1$, they are Killing circles.

Properties of extrinsic shapes of trajectories are obtained by direct computations. If we denote by ∇ and $\widetilde{\nabla}$ the covariant differentiations on M and $\widetilde{M}$, respectively, we have the Gauss formula $\widetilde{\nabla}_X Y = \nabla_X Y + \langle AX, Y \rangle \mathcal{N}$ and the Weingarten formula $\widetilde{\nabla}_X \mathcal{N} = -AX$ for arbitrary vector fields X, Y on M. Since we have $A\dot{\gamma} = \lambda_M \dot{\gamma} + (\nu_M - \lambda_M)\rho_\gamma \xi$ for a trajectory γ for $\mathbb{F}_\kappa$ on a hypersurface M of type either (A_0) or (A_1), we obtain the following.

Proposition 7.1 ([3]). *The extrinsic shape of a trajectory γ for $\mathbb{F}_\kappa$ on a hypersurface M of type either* (A_0) *or* (A_1) *in $\mathbb{C}H^n(-c)$ lies on some totally geodesic $\mathbb{C}H^2$. Its property is as follows:*

1) *When $\kappa = 0$ and $\lambda_M + (\nu_M - \lambda_M)\rho_\gamma^2 = 0$, it is a geodesic;*
2) *When $\lambda_M + (\nu_M - \lambda_M)\rho_\gamma^2 = \kappa \rho_\gamma$, it is a trajectory for a Kähler magnetic field $\mathbb{B}_\kappa$, hence is a circle of positive geodesic curvature except the case* 1);
3) *When $\kappa = (\lambda_M - \nu_M)\rho_\gamma$, it is a circle of positive geodesic curvature with*

$$k_1 = \sqrt{\kappa^2(1-\rho_\gamma^2) + \{\lambda_M + (\nu_M - \lambda_M)\rho_\gamma^2\}^2},$$

$$\tau_{12} = -\{\lambda_M \rho_\gamma + (\nu_M - \lambda_M)\rho_\gamma^3 + (1-\rho_\gamma^2)\kappa\} \,/\, k_1;$$

4) *When $\kappa \nu_M \rho_\gamma = \lambda_M^2 + \lambda_M(\nu_M - \lambda_M)\rho_\gamma^2$, it is a Killing helix of proper order 3;*
5) *Otherwise, it is a Killing helix of proper order 4.*

If we restrict ourselves on circular trajectories on hypersurfaces of types (A_0) and (A_1), as they satisfy $\kappa \rho_\gamma = \lambda$, we obtain the following.

Corollary 7.1. *The extrinsic shape of a circular trajectory γ for $\mathbb{F}_\kappa$ on a hypersurface M of type either* (A_0) *or* (A_1) *in $\mathbb{C}H^n(-c)$ is a Killing helix of proper order 4. Its geodesic curvatures k_1, k_2, k_3 are*

$$\frac{1}{\kappa^2}\sqrt{\kappa^6 + (1+2\kappa^2)\lambda_M^2}, \quad \frac{(\kappa^2+1)\lambda_M\sqrt{\kappa^2-\lambda_M^2}}{\kappa^2\sqrt{\kappa^6+(1+2\kappa^2)\lambda_M^2}}, \quad \frac{\kappa^2-\lambda_M^2}{\sqrt{\kappa^6+(1+2\kappa^2)\lambda_M^2}},$$

and its complex torsions satisfy $\tau_{13} = \tau_{24} = 0$ and

$$\tau_{12} = \tau_{34} = \frac{-\mathrm{sgn}(\kappa) \cdot (k_1+k_3)}{\sqrt{k_2^2 + (k_1+k_3)^2}}, \quad \tau_{23} = \tau_{14} = \frac{-\mathrm{sgn}(\kappa) \cdot k_2}{\sqrt{k_2^2 + (k_1+k_3)^2}}.$$

The above Corollary shows that the extrinsic shape of each circular trajectory lies on some totally geodesic $\mathbb{C}H^2$. Such a helix of proper order 4 on $\mathbb{C}H^n$ is called an essential Killing helix (see [4]).

8. Asymptotic behaviors of circular trajectories

In this section and next we consider properties of circular trajectories. Complex hyperbolic spaces are simply connected and of negative curvature. They are hence examples of Hadamard manifolds. For a Hadamard manifold M we can consider its ideal boundary ∂M and a compactification $\overline{M} = M \cup \partial M$ with the cone topology (see [13]). For a complex hyperbolic space $\mathbb{C}H^n$ we have its ball model

$$D^n = \left\{ w = (w_1, \ldots, w_n) \in \mathbb{C}^n \mid |w_1|^2 + \cdots + |w_n|^2 < 1 \right\}.$$

Let $\varpi : H_1^{2n+1} \to \mathbb{C}H^n$ denote a fibration of an anti-de Sitter space

$$H_1^{2n+1} = \left\{ z = (z_0, \ldots, z_n) \in \mathbb{C}^{n+1} \mid -|z_0|^2 + |z_1|^2 + \cdots + |z_n|^2 = -1 \right\}.$$

If we take $z = (z_0, \ldots, z_n) \in H_1^{2n+1} \subset \mathbb{C}^{n+1}$, the point $\varpi(z) \in \mathbb{C}H^n$ corresponds to the point $(z_1/z_0, \ldots, z_n/z_0) \in D^n$. Under this identification the ideal boundary of $\mathbb{C}H^n$ corresponds to the topological boundary of D^n.

We call a smooth curve σ on $\mathbb{C}H^n$ parameterized by its arclength unbounded in both directions if both of the sets $\sigma([0, \infty))$ and $\sigma((-\infty, 0])$ are unbounded. For a smooth curve on $\mathbb{C}H^n$ which is unbounded in both directions we put $\sigma(\infty) = \lim_{t \to \infty} \sigma(t)$, $\sigma(-\infty) = \lim_{t \to -\infty} \sigma(t)$ if they exist in $\partial \mathbb{C}H^n$. We call them *points at infinity* of σ. Since horospheres and tubes around totally geodesic $\mathbb{C}H^{n-1}$ are unbounded sets in $\mathbb{C}H^n$, for curves on these real hypersurfaces, by considering their extrinsic shapes we use these terminologies.

For the sake of simplicity, we study in $\mathbb{C}H^n(-4)$. Let γ be a circular trajectory on a real hypersurface M which is congruent to one of $HS, G(r)$ and $T(r)$ in $\mathbb{C}H^n(-4)$. We take a horizontal lift $\hat{\gamma}$ of the extrinsic shape of γ with respect to a fibration $\varpi : H_1^{2n+1} \to \mathbb{C}H^n(-4)$. We can regard this curve $\hat{\gamma}$ as a curve in $\mathbb{C}^{n+1}$. Riemannian connections $\overline{\nabla}$ and $\widetilde{\nabla}$ on $\mathbb{C}^{n+1}$ and on $\mathbb{C}H^n(-4)$ are related as

$$\overline{\nabla}_{\widetilde{X}} \widetilde{Y} = \widetilde{\nabla}_{\widetilde{X}} \widetilde{Y} + \langle \widetilde{X}, \widetilde{Y} \rangle \widehat{N} - \langle \widetilde{X}, J\widetilde{Y} \rangle J\widehat{N}$$

with normal vector field $\widehat{N}$ on H_1^{2n+1} satisfying $\langle \widehat{N}, \widehat{N} \rangle = -1$, where arbitrary vector fields $\widetilde{X}, \widetilde{Y}$ on $\mathbb{C}H^n(-4)$ are regarded as horizontal vector fields on H_1^{2n+1}. Thus we find by Corollary 7.1 that $\hat{\gamma}$ satisfies the following differential equation

$$\hat{\gamma}''' - \sqrt{-1}(\kappa + \kappa^{-1})\hat{\gamma}'' - (2 - \rho_\gamma^2)\hat{\gamma}' + \sqrt{-1}(1 - \rho_\gamma^2)\kappa^{-1}\hat{\gamma} = 0.$$

It is clear that γ is not bounded if and only if the characteristic equation

$$\Lambda^3 - \sqrt{-1}(\kappa + \kappa^{-1})\Lambda^2 - (2 - \lambda_M^2 \kappa^{-2})\Lambda + \sqrt{-1}(1 - \lambda_M^2 \kappa^{-2})\kappa^{-1} = 0 \quad (5)$$

of this differential equation has a double solution or a solution which is not pure imaginary. When M is a horosphere HS, as $\lambda_{HS} = 1$, the characteristic equation (5) has a double solution. We hence get the following:

Theorem 8.1 ([11]). *Every circular trajectory for a Sasakian magnetic field on a horosphere HS in $\mathbb{C}H^n$ is unbounded in both directions, hence has single point of infinity.*

When M is a tube $T(r)$ around totally geodesic $\mathbb{C}H^{n-1}$, the characteristic equation (5) turns to

$$\Omega^3 - \frac{1}{3}\left\{\kappa^2 - 4 + (1 + 3\tanh^2 r)\kappa^{-2}\right\}\Omega$$
$$- \frac{1}{27}\left\{2\kappa^3 - 12\kappa + 3(5 + 3\tanh^2 r)\kappa^{-1} + 2(1 - 9\tanh^2 r)\kappa^{-3}\right\} = 0, \tag{6}$$

if we make a parallel translation $\Omega = -\sqrt{-1}\Lambda - (\kappa + \kappa^{-1})/3$. When the coefficient $\kappa^2 - 4 + (1 + 3\tanh^2 r)\kappa^{-2}$ of Ω^2 is not positive, the equation (6) has only one real solution, which means that the original characteristic equation has solutions which are not pure imaginary. When the coefficient of Ω^2 is positive, the equation (6) turns to

$$\theta^3 - (3/2)\theta + \tau_T(\kappa; r)/\sqrt{2} = 0, \tag{7}$$

with

$$\tau_T(\kappa; r) = -\operatorname{sgn}(\kappa)\frac{(\kappa^2 - 2)(2\kappa^4 - 8\kappa^2 + 9\tanh^2 r - 1)}{2(\kappa^4 - 4\kappa^2 + 3\tanh^2 r + 1)^{3/2}},$$

if we make a homothetic change of the variable. This equation has three distinct real solutions if and only if $|\tau_T(\kappa; r)| < 1$. We hence obtain the following:

Theorem 8.2 ([11]). *We consider a circular trajectory γ for $\mathbb{F}_\kappa$ on a tube $T(r)$ in $\mathbb{C}H^n(-4)$.*

(1) *If κ satisfies $2\{1 - (\cosh r)^{-1}\} \le \kappa^2 \le 2\{1 + (\cosh r)^{-1}\}$, it is unbounded in both directions. When one of the equalities holds, it has single point at infinity. Otherwise it has two distinct points at infinity.*
(2) *If κ satisfies either $\tanh^2 r < \kappa^2 < 2\{1 - (\cosh r)^{-1}\}$ or $\kappa^2 > 2\{1 + (\cosh r)^{-1}\}$, then it is bounded.*

We here note that behavior of geodesics, which can be said as trajectories for the trivial magnetic field $\mathbb{F}_0$, on hypersurfaces of types (A$_0$) and (A$_1$) were studied in [7].

9. Lengths of circular trajectories

In this section we study a condition for circular trajectories to be closed on a real hypersurface which is congruent to either a geodesic sphere or a tube around totally geodesic $\mathbb{C}H^{n-1}$ in $\mathbb{C}H^n$. A smooth curve σ parameterized by its arclength is said to be *closed* if there is a positive t_0 satisfying $\sigma(t + t_0) = \sigma(t)$ for all t. The minimum positive t_0 with this property is called the *length* of closed curve σ and is denoted by $\text{length}(\sigma)$. If a smooth curve σ is not closed we say it is open and set $\text{length}(\sigma) = \infty$.

We make use of the notations in the previous section. If the characteristic equation (5) for a horizontal lift of a trajectory γ has three distinct pure imaginary solutions $\sqrt{-1}a_\kappa, \sqrt{-1}b_\kappa, \sqrt{-1}c_\kappa$ $(a_\kappa < b_\kappa < c_\kappa)$, then the extrinsic shape $\tilde{\gamma}$ of γ is of the form $\tilde{\gamma}(t) = \varpi\big(Ae^{\sqrt{-1}a_\kappa t} + Be^{\sqrt{-1}b_\kappa t} + Ce^{\sqrt{-1}c_\kappa t}\big)$ with $\mathbb{C}$-linearly independent $A, B, C \in \mathbb{C}^{n+1}$. Therefore we find that γ is closed if and only if there exists a constant d satisfying that the ratios $(a_\kappa - d)/(b_\kappa - d)$, $(b_\kappa - d)/(c_\kappa - d)$ are rational and that in this case its length is given as $2\pi \times \text{L.C.M.}\{(b_\kappa - a_\kappa)^{-1}, (c_\kappa - a_\kappa)^{-1}\}$. Here $\text{L.C.M.}(\alpha, \beta)$ for positive numbers α, β denotes the minimum positive number in the set $\{j\alpha \mid j = 1, 2, \ldots\} \cap \{j\beta \mid j = 1, 2, \ldots\}$. Since the equation (7) is obtained by a parallel translation and a homothetic change of variables from (5), we find that there exists a constant d satisfying the ratios $(a_\kappa - d)/(b_\kappa - d)$, $(b_\kappa - d)/(c_\kappa - d)$ are rational if and only if the solutions of the equation (7) have this property. As we see the equation (7) coincides with the realized characteristic equation $\theta^3 - (3/2)\theta + \tau/\sqrt{2} = 0$ for the differential equation $\hat{\sigma}''' + (3/2)\hat{\sigma}' - \sqrt{-1}(\tau/\sqrt{2})\hat{\sigma} = 0$ of a horizontal lift $\hat{\sigma}$ of a circle σ of geodesic curvature $1/\sqrt{2}$ and complex torsion τ on $\mathbb{C}P^n(4)$, we can transplant properties of circles on $\mathbb{C}P^n(4)$ (see [6]) to our circular trajectories.

Theorem 9.1 ([11]). *We consider circular trajectory γ for $\mathbb{F}_\kappa$ on $T(r)$ in $\mathbb{C}H^n(-4)$.*

(1) *When $r \leq \log(\sqrt{2}+1)$ and $\kappa^2 = \{4 + 3\sqrt{2}(\cosh r)^{-1}\}/2$, it is closed of length $2\pi\sqrt{\cosh r\,(4\cosh r + 3\sqrt{2})}$.*

(2) *When $r > \log(\sqrt{2}+1)$ and $\kappa^2 = \{4 \pm 3\sqrt{2}(\cosh r)^{-1}\}/2$, it is closed of length $2\pi\sqrt{\cosh r\,(4\cosh r \pm 3\sqrt{2})}$, where double signs take the same signatures.*

(3) *If κ satisfies either $\tanh^2 r < \kappa^2 < 2\{1 - (\cosh r)^{-1}\}$ or $\kappa^2 > 2\{1 + (\cosh r)^{-1}\}$ and is not in the cases of (1) and (2), it is closed if and*

only if

$$\frac{\left|(\kappa^2-2)(2\kappa^4-8\kappa^2+9\tanh^2 r-1)\right|}{2(\kappa^4-4\kappa^2+3\tanh^2 r+1)^{3/2}} = \frac{q(9p^2-q^2)}{(3p^2+q^2)^{3/2}}$$

holds with some relatively prime positive integers p, q *satisfying* $p > q$. *In this case its length is given as*

$$\pi\delta(p,q)|\kappa|\sqrt{(3p^2+q^2)/(\kappa^4-4\kappa^2+3\tanh^2 r+1)},$$

where $\delta(p,q) = 1$ *when* pq *is odd and* $\delta(p,q) = 2$ *when* pq *is even.*

When M is a geodesic sphere $G(r)$ in $\mathbb{C}H^n(-4)$, if we make a parallel translation and a homothetic change of variable, we find the characteristic equation (5) with $\lambda_{G(r)} = \cosh r$ turns to $\theta^3 - (3/2)\theta + \tau_G(\kappa; r)/\sqrt{2} = 0$ with

$$\tau_G(\kappa; r) = -\mathrm{sgn}(\kappa)\frac{(\kappa^2-2)(2\kappa^4-8\kappa^2+9\coth^2 r-1)}{2(\kappa^4-4\kappa^2+3\coth^2 r+1)^{3/2}}.$$

In this case also we can transplant properties of circles on $\mathbb{C}P^n(4)$.

Theorem 9.2 ([11]). *We consider circular trajectory* γ *for* $\mathbb{F}_\kappa$ *on a geodesic sphere* $G(r)$ *in* $\mathbb{C}H^n(-4)$.

(1) *When* $r > \log(\sqrt{2}+1)$ *and* $\kappa = \pm\sqrt{2}$, *it is closed and its length is* $2\sqrt{2}\pi\sinh r$.

(2) *Otherwise, it is closed if and only if*

$$\frac{\left|\kappa^2-2\right|(2\kappa^4-8\kappa^2+9\coth^2 r-1)}{2(\kappa^4-4\kappa^2+3\coth^2 r+1)^{3/2}} = \frac{q(9p^2-q^2)}{(3p^2+q^2)^{3/2}}$$

holds with some relatively prime positive integers p, q *satisfying* $p > q$. *In this case its length is given as*

$$\pi\delta(p,q)|\kappa|\sqrt{(3p^2+q^2)/(\kappa^4-4\kappa^2+3\coth^2 r+1)},$$

where $\delta(p,q) = 1$ *when* pq *is odd and* $\delta(p,q) = 2$ *when* pq *is even.*

We are now interested in how lengths of circular trajectories are distributed. We say two smooth curves σ_1, σ_2 parameterized by their arclengths on a Riemannian manifold N *congruent* to each other if there exist an isometry φ of M and a constant t_c satisfying $\sigma_2(t+t_c) = \varphi \circ \sigma_1(t)$ for all t. For trajectories for Sasakian magnetic fields on a hypersurface of type (A) we have the following congruence theorems. These also show the importance of structure torsions and principal torsions for trajectories on hypersurfaces of type (A).

Proposition 9.1 ([3]). *Let M be a hypersurface of type either (A_0) or (A_1) in $\mathbb{C}H^n$. Trajectories γ_1 for a Sasakian magnetic field $\mathbb{F}_{\kappa_1}$ and γ_2 for $\mathbb{F}_{\kappa_2}$ on M are congruent to each other if and only if one of the following conditions holds:*

i) $|\rho_{\gamma_1}| = |\rho_{\gamma_2}| = 1$,
ii) $\rho_{\gamma_1} = \rho_{\gamma_2} = 0$ *and* $|\kappa_1| = |\kappa_2|$,
iii) $0 < |\rho_{\gamma_1}| = |\rho_{\gamma_2}| < 1$ *and* $\kappa_1\rho_{\gamma_1} = \kappa_2\rho_{\gamma_2}$.

Proposition 9.2 ([8]). *Let M be a hypersurface of type (A_2) in $\mathbb{C}H^n$. Trajectories γ_1 for a Sasakian magnetic field $\mathbb{F}_{\kappa_1}$ and γ_2 for $\mathbb{F}_{\kappa_2}$ on M are congruent to each other if and only if one of the following conditions holds:*

i) $|\rho_{\gamma_1}| = |\rho_{\gamma_2}| = 1$,
ii) $\rho_{\gamma_1} = \rho_{\gamma_2} = 0$, $\tau_{\gamma_1} = \tau_{\gamma_2}$ *and* $|\kappa_1| = |\kappa_2|$,
iii) $0 < |\rho_{\gamma_1}| = |\rho_{\gamma_2}| < 1$, $\tau_{\gamma_1} = \tau_{\gamma_2}$ *and* $\kappa_1\rho_{\gamma_1} = \kappa_2\rho_{\gamma_2}$.

For a real hypersurface M of type (A) in $\mathbb{C}H^n$ we denote by $\mathcal{T}_\phi(M)$ the set of all congruence classes of circular trajectories. When M is congruent to either $G(r)$ or $T(r)$, by Corollaries 5.2 and 5.3, we see it is set theoretically bijective to an open interval (λ_M, ∞). We define $\mathcal{L}_c : \mathcal{T}_\phi(M) \to (0, \infty]$ by $\mathcal{L}_c([\gamma]) = \text{length}(\gamma)$, where $[\gamma]$ denotes the congruence class containing a circular trajectory γ. We call $\mathcal{L}_c$ the *length spectrum* of circular trajectories. We also call the set $\text{LSpec}_\phi(M) = \mathcal{L}_c(\mathcal{T}_\phi(M)) \cap \mathbb{R}$ the length spectrum of circular trajectories. For a real numbe λ, we call the cardinality of the set $\mathcal{L}_c^{-1}(\lambda)$ *multiplicity* of the length spectrum at λ.

Theorem 9.3 ([11]). *Let M be a real hypersurface which is congruent to either a geodesic sphere $G(r)$ or a tube $T(r)$ around totally geodesic $\mathbb{C}H^{n-1}$ in $\mathbb{C}H^n$.*

(1) *The length spectrum $\text{LSpec}_\phi(M)$ of circular trajectories on M is an unbounded discrete subset in $\mathbb{R}$.*
(2) *At each $\lambda \in \text{LSpec}_\phi(M)$, its multiplicity is finite.*

When the multiplicity is one, we say this length simple, we can distinguish a congruence class of circular trajectories having this length only by their lengths. We are interested in the behavior of multiplicities.

References

1. T. Adachi, *Kähler magnetic flows on a manifold of constant holomorphic sectional curvature*, Tokyo J. Math. 18(1995), 473–483.

2. _____, *Distribution of length spectrum of circles on a complex hyperbolic space*, Nagoya Math. J. 153(1999), 119–140.

3. _____, *Trajectories on geodesic spheres in a non-flat complex space form.* J. Geom. 90(2008), 1–29.

4. _____, *Essential Killing helices of order less than five on a non-flat complex space form*, to appear in J. Math. Soc. Japan.

5. _____, *Magnetic Jacobi fields for Kähler magnetic fields*, in this volume.

6. T. Adachi, S. Maeda and S. Udagawa, *Circles in a complex projective space*, Osaka J. Math. 32(1995), 709–719.

7. T. Adachi, S. Maeda and M. Yamagishi, *Length spectrum of geodesic spheres in a non-flat complex space form*, J. Math. Soc. Japan 54(2002), 373–408.

8. T. Bao and T. Adachi, *Circular trajectories on real hypersurfaces in a nonflat complex space form*, J. Geom. 96(2009), 41–55.

9. _____, *Lengths of circular trajectories on geodesic spheres in a complex projective space*, Kodai Math. J. 34(2011), 257–271.

10. _____, *Trajectories for Sasakian magnetic fields on homogeneous real hypersurfaces of type* (B) *in a complex hyperbolic space*, to appear in Diff. Geom. Appl., doi:10.1016/j.difgeo.2011.04.004.

11. _____, *Behavior of circular trajectories on hypersurfaces of type* (A_1) *in a complex hyperbolic space*, to appear in Kodai Math. J.

12. J. Berndt, *Real hypersurfaces with constant principal curvatures in complex hyperbolic space*, J. Reine Angew. Math. 395(1989), 132–141.

13. P. Eberlein and B. O'Neill, *Visibility manifolds*, Pasific J. Math. 46(1973), 45–109.

14. U-H. Ki, H.S. Kim and H. Nakagawa, *A characterization of a real hypersurface of type* (B), Tsukuba J. Math. 14(1980), 9–26.

15. S. Maeda and T. Adachi, *Sasakian curves on hypersurfaces of type* (A) *in a nonflat complex space form*, Results in Math. 56(2009), 489–499.

16. S. Maeda and Y. Ohnita, *Helical geodesic immersions into complex space form*, Geom. Dedicata 30(1989), 93–114.

17. R. Niebeergall and P.J. Ryan, *Real hypersurfaces in complex space forms*, Tight and Taut Submanifolds, T.E. Cecil and S.S. Chern eds., Cambridge Umiv. Press, Cambridge (1998), 233–305.

18. T. Sunada, *Magnetic flows on a Riemann surface*, Proc. KAIST Math. Workshop 8(1993), 93–108.

Received January 23, 2011
Revised   April 25, 2011

Proceedings of the 2nd International
Colloquium on Differential Geometry
and its Related Fields
Veliko Tarnovo, September 6–10, 2010

TYZ EXPANSIONS FOR SOME ROTATION
INVARIANT KÄHLER METRICS

Todor GRAMCHEV* and Andrea LOI**

*Dipartimento di Matematica e Informatica, Università di Cagliari,
Cagliari, 09124, Italy
*E-mail: todor@unica.it
**E-mail: loi@unica.it
www.unica.it*

The main goal of the paper is to make a survey and outline some new results on the existence of Kempf's distortion function and the Tian-Yau-Zelditch (TYZ) asymptotic expansion for noncompact Kähler manifolds in two important rotation invariant cases: the punctured plane $(\mathbb{C}^*, g^*)$ with the complete cylindrical metric and the Kepler manifold (X, g).

Keywords: Kähler manifolds, Quantization, Quantum mechanics, TYZ asymptotic expansion, Exponential reminder.

1. Introduction

We start by recalling some basic geometric facts in the theory of Tian–Yau–Zelditch (TYZ) asymptotic expansions, following our previous work [1] (see also [2, 3, 5, 6, 7, 8, 9, 10, 11, 12, 13, 14, 15, 16, 17, 18, 19]).

Let g be a Kähler metric on an n–dimensional complex manifold M. We assume that g is polarized with respect to a holomorphic line bundle L over M, i.e.

$$c_1(L) = [\omega],$$

where ω stands for the Kähler form associated to g and $c_1(L)$ denotes the first Chern class of L.

Let $m \in \mathbb{N}$. We denote by h_m the Hermitian metric on $L^m = L^{\otimes m}$ such that its Ricci curvature

$$\mathrm{Ric}(h_m) = m\omega,$$

with $\mathrm{Ric}(h_m)$ being the two–form on M whose local expression is given by

$$\mathrm{Ric}(h_m) = -\frac{i}{2}\partial\bar{\partial}\log h_m\big(\sigma(x), \sigma(x)\big), \tag{1}$$

for a trivializing holomorphic section

$$\sigma : U \to L^m \setminus \{0\}.$$

In the quantum mechanics terminology L^m is called the *quantum line bundle*, the pair (L^m, h_m) is called a *geometric quantization* of the Kähler manifold $(M, m\omega)$ and $\hbar = m^{-1}$ plays the role of Planck's constant (e.g., see [2]). Consider the separable complex Hilbert space $\mathcal{H}_m$ defined as the set of all global holomorphic sections s of L^m such that

$$\langle s, s \rangle_m = \int_M h_m(s(x), s(x)) \frac{\omega^n}{n!} < \infty.$$

This Hilbert space can be trivial in general, i.e. $\mathcal{H}_m = \{0\}$, whereas it is infinite dimensional for the two cases discussed in the paper. More precisely, in the rest of the paper we will assume $\mathcal{H}_m \neq \{0\}$, condition satisfied by the punctured plane and the Kepler manifold, where $\mathcal{H}_m$ is indeed infinite dimensional.

Let $x \in M$ and let $q \in L^m \setminus \{0\}$ be a fixed point of the fiber over x. If one evaluates $s \in \mathcal{H}_m$ at x, one gets a multiple $\delta_q(s)$ of q, i.e. $s(x) = \delta_q(s)q$. The map $\delta_q : \mathcal{H}_m \to \mathbb{C}$ is a continuous linear functional ([27]). Hence from Riesz's theorem, there exists a unique $e_q^m \in \mathcal{H}$ such that

$$\delta_q(s) = \langle s, e_q^m \rangle_m, \forall s \in \mathcal{H}_m,$$

i.e.

$$s(x) = \langle s, e_q^m \rangle_m q. \tag{2}$$

It follows that $e_{cq}^m = \overline{c}^{-1} e_q^m$, $\forall c \in \mathbb{C}^*$. The holomorphic section $e_q^m \in \mathcal{H}_m$ is called the *coherent state* relative to the point q. Thus, one can define a smooth function on M

$$T_m(x) = h_m(q, q) \|e_q^m\|^2, \quad \|e_q^m\|^2 = \langle e_q^m, e_q^m \rangle, \tag{3}$$

where $q \in L^m \setminus \{0\}$ is any point on the fiber of x. If s_j, $j = 0, \ldots d_m$, $(d_m + 1 = \dim \mathcal{H}_m \leq \infty)$ form an orthonormal basis for $(\mathcal{H}_m, \langle \cdot, \cdot \rangle_m)$ then one can easily verify that

$$T_m(x) = \sum_{j=0}^{d_m} h_m(s_j(x), s_j(x)). \tag{4}$$

We point out that the dimension $d_m + 1$ is finite and hence (4) is a finite sum when M is compact. Indeed, $\mathcal{H}_m = H^0(L^m)$, where $H^0(L^m)$ denotes the space of global holomorphic sections of L^m. The function T_m

is *Kempf's distortion function*. We refer the reader to the paper [1] and references therein for more details on Kempf distortion function.

Choose and fix a positive integer m. Under the hypothesis that for each point $x \in M$ there exists $s \in \mathcal{H}_m$ non-vanishing at x, we can give the following geometric interpretation of T_m. Consider the holomorphic map of M into the complex projective space $\mathbb{CP}^{d_m}$:

$$\varphi_m : M \to \mathbb{CP}^{d_m} : x \mapsto [s_0(x) : \cdots : s_{d_m}(x)]. \tag{5}$$

One can prove that

$$\varphi_m^*(\omega_{FS}) = m\omega + \frac{i}{2}\partial\bar{\partial}\log T_m, \tag{6}$$

where ω_{FS} is the Fubini–Study form on $\mathbb{CP}^{d_m}$, namely the form which in homogeneous coordinates $[Z_0, \ldots, Z_{d_m}]$ reads as $\omega_{FS} = \frac{i}{2}\partial\bar{\partial}\log\sum_{j=0}^{d_m}|Z_j|^2$. Clearly (6) leads to

$$\frac{\varphi_m^*(\omega_{FS})}{m} - \omega = \frac{i}{2m}\partial\bar{\partial}\log T_m. \tag{7}$$

Hence, the term

$$\mathcal{E}_m(x) := \frac{i}{2m}\partial\bar{\partial}R_m, \tag{8}$$

where $R_m = \log T_m$ turns out to play a role of the "error" of the approximation of ω (resp. g) by $\frac{\varphi_m^*(\omega_{FS})}{m}$ (resp. $\frac{\varphi_m^*(g_{FS})}{m}$).

We point out that that, by (6), if mg is a balanced metric, or more generally if T_m is harmonic, for some $m \in \mathbb{N}$, then $\mathcal{E}_m(x) = 0$ and hence mg is projectively induced via the coherent states map φ_m (see [2, 37, 38] for more details on the link between balanced metrics and quantization of Kähler manifolds).

Recall that a Kähler metric g on a complex manifold M is *projectively induced* if there exists a Kähler (i.e. a holomorphic and isometric) immersion

$$\psi : M \to \mathbb{CP}^N, \qquad N \leq \infty,$$

such that $\psi^*(g_{FS}) = g$. Projectively induced Kähler metrics enjoy important geometrical properties and were extensively studied in [21] (see also [11]).

Not all Kähler metrics are balanced or projectively induced.

In the case of compact manifold M, Tian[17] and Ruan[16] solved a conjecture posed by Yau by proving that the sequence of metrics $\frac{\varphi_m^*(\omega_{FS})}{m}$ converges to ω in the C^∞ category, i.e., any polarized metric on compact complex manifold is the C^∞-limit of (normalized) projectively induced Kähler metrics. Zelditch[18] generalized the Tian–Ruan theorem by proving a complete asymptotic expansion in the C^∞ category, namely

$$T_m(x) \sim \sum_{j=0}^{\infty} a_j(x) m^{n-j}, \qquad (9)$$

where a_j, $j = 0, 1, \ldots$, are smooth coefficients with $a_0(x) = 1$, and for any nonnegative integers r, k we can find positive constant C depending on n k, r and on the Kähler form ω such that the following estimates hold:

$$\left\| T_m(x) - \sum_{j=0}^{k} a_j(x) m^{n-j} \right\|_{C^r} \leq C \, m^{n-k-1}, \qquad m \geq 1 \qquad (10)$$

where $\| \cdot \|_{C^r}$ denotes the C^r norm in local coordinates. We point out that similar asymptotic expansion were used in [27, 28, 29, 30, 31, 32] to construct a star product on Kähler manifolds.

Later on, Lu [20], by means of Tian's peak section method, proved that each of the coefficients $a_j(x)$ in (9) is a polynomial of the curvature and its covariant derivatives at x of the metric g. Such polynomials can be found by finitely many algebraic operations. Furthermore $a_1(x) = \frac{1}{2}\rho$, where ρ is the scalar curvature of the polarized metric g (see also [7] and [8] for the computations of the coefficients a_j's through Calabi's diastasis function). The expansion (9) is called the *TYZ* (*Tian–Yau–Zelditch*) *expansion* and is a key ingredient in the investigations of balanced metrics ([37]).

We are inspired by the investigations of geometric quantization problems cf. [24, 25, 26] where analytical tools have been applied in order to extend Berezin's quantization method (cf. [22, 23]) to non homogeneous complex domains on $\mathbb{C}^n$ (see also [33, 34, 35] and the references therein for the study of coherent states and relations to geometric quantization).

In our previous paper [1] we have computed explicitly the Kempf distortion function $T_m(x)$ for the Kepler manifold (X, ω), using the rotation symmetry representation. Using this computation we derived an analogue of Zelditch and Lu's theorems above for the Kepler manifold (X, ω).

The goal of the present work is to give a survey and propose refinements and some new results on TYZ asymptotic expansions on non compact Kähler manifolds whose Kähler form is either $U(1)$-invariant (the punctured plane) or $O(n)$-invariant (the Kepler manifold).

The paper is organized as follows. Section 2 deals with the Bergmann metric on the punctured complex plane C^*. Here we propose a generalization of the convergence results and estimates of the error term R_m. In Section 3 we propose, following our work [1], an explicit construction of the

Kempf distortion function T_m for the Kepler manifold (X, ω), with some refinements of the representations. The last section outlines the main result in [1], i.e. an exact TYZ asymptotic expansion and the exponentially small decay for the remainder when $m \to \infty$. Here we propose some functional analytic improvements.

2. On the remainder term for the cylindrical metric on $\mathbb{C}^*$

In this section we deal with the case of the punctured plane $\mathbb{C}^* = \mathbb{C} \setminus \{0\}$ equipped with the complete Kähler metric g^* whose associated Kähler form is given by

$$\omega^* = \frac{i}{2} \frac{dz \wedge d\bar{z}}{|z|^2}$$

and the polarization L is given by the trivial bundle $L = \mathbb{C}^* \times \mathbb{C}$ and hence for all natural numbers m, $L^m = \mathbb{C}^* \times \mathbb{C}$. Let h_m be the hermitian metric on L^m given by:

$$h_m(f(z), f(z)) := e^{\frac{-m}{2} \log^2 |z|^2} |f(z)|^2$$

for a holomorphic function f on $\mathbb{C}^*$. It is easily seen that $\text{Ric}(h_m) = m\omega^*$ and hence L^m is a quantization of $(\mathbb{C}^*, mg^*)$. It follows that the space $\mathcal{H}_m$, equals the space of holomorphic functions f in $\mathbb{C}^*$ such that

$$\|f\|_m^2 = \langle f, f \rangle_m = \int_{\mathbb{C}^*} e^{\frac{-m}{2} \log^2 |z|^2} |f(z)|^2 \frac{i}{2} \frac{dz \wedge d\bar{z}}{|z|^2} < +\infty.$$

One can check that the functions z^j, with $j \in \mathbb{Z}$, form an orthogonal system for $\mathcal{H}_m$. Since every holomorphic function in $\mathbb{C}^*$ can be expanded in Laurent series, it follows that z^j are in fact a complete orthogonal system. Their norms are given by

$$\|z^j\|_m^2 = \int_{\mathbb{C}^*} e^{\frac{-m}{2} \log^2 |z|^2} |z|^{2j} \frac{i}{2} \frac{dz \wedge d\bar{z}}{|z|^2}$$

$$= 2\pi \int_0^{+\infty} e^{\frac{-m}{2} \log^2 r^2} r^{2j} \frac{r}{r^2} \, dr.$$

By the change of variable $e^\rho = r^2$ one gets

$$\|z^j\|_m^2 = \pi \int_{-\infty}^{+\infty} e^{j\rho - \frac{m}{2}\rho^2} d\rho = \pi e^{\frac{j^2}{2m}} \int_{-\infty}^{+\infty} e^{-\left(\sqrt{\frac{m}{2}}\rho - \sqrt{\frac{1}{2m}}j\right)^2} d\rho$$

$$= \pi e^{\frac{j^2}{2m}} \sqrt{\frac{2}{m}} \int_{-\infty}^{+\infty} e^{-t^2} dt = \sqrt{\frac{2\pi^3}{m}} \, e^{\frac{j^2}{2m}}.$$

Then, an orthonormal basis for $\mathcal{H}_m$ is given by

$$s_j = \left(\sqrt{\frac{m}{2\pi^3}}e^{-\frac{j^2}{2m}}\right)^{\frac{1}{2}} z^j$$

and

$$\varphi_m^*(\omega_{FS}) = \omega_m = \frac{i}{2}\partial\bar{\partial}\log\sum_{j=0}^{+\infty}|s_j|^2 = \frac{i}{2}\partial\bar{\partial}\log\sum_{j\in\mathbb{Z}}e^{-\frac{j^2}{2m}}|z|^{2j}, \qquad (11)$$

where φ_m is given by (5).

Let $\frac{g_m}{m}$ be the corresponding sequence of Bergmann metrics. The following theorem represents the main result in [6]. We provide here a proof for completeness.

Theorem 2.1. *Let $\mathbb{C}^*$ be endowed with the complete metric g^*. Then the sequence of Bergmann metrics $\frac{g_m}{m}$ C^∞-converges to the metric g^* on every compact set $K \subset \mathbb{C}^*$.*

Proof. By formula (11) it is enough to show that the sequence of functions

$$f_m(t) = \frac{1}{m}\log\left(\sum_{j\in\mathbb{Z}}f_m(j;t)\right), \quad m \in \mathbb{N}, \ t > 0, \qquad (12)$$

C^∞-converges to the function $f(t) = \frac{1}{2}\log^2 t$ on every compact set $C \subset \mathbb{R}^+$, where

$$f_m(j;t) = \exp\left(-\frac{j^2}{2m}\right)t^j = \exp\left(-\frac{j^2}{2m} + j\log t\right), \quad j \in \mathbb{Z}, t > 0. \quad (13)$$

In order to prove it, we apply the Poisson summation formula (see p. 347, Theorem 24 in [40]) to the functions $f_m(j;t)$, $j \in \mathbb{Z}$, namely

$$\sum_{j\in\mathbb{Z}}f_m(j;t) = \sum_{j\in\mathbb{Z}}\widehat{f_m}(j;t), \qquad (14)$$

with

$$\widehat{f_m}(j;t) = \int_{-\infty}^{\infty}\exp(-2\pi ijs)f_m(s;t)\,ds. \qquad (15)$$

By (14) and direct calculations, we obtain that

$$\widehat{f_m}(j;t) = \int_{-\infty}^{\infty}\exp\left(-2\pi ijs - \frac{s^2}{2m} + s\log t\right)ds$$

$$= \sqrt{2\pi m}\exp\left(\frac{m}{2}\log^2 t\right)\exp(-2\pi^2 mj^2 - i2\pi mj\log t). \quad (16)$$

Next, taking into account (16), we write in a new form $f_m(t)$ defined in (22)

$$
\begin{aligned}
f_m(t) &= \frac{1}{m} \log\left(\sqrt{2\pi m} \exp(\frac{m}{2} \log^2 t) \sum_{j \in \mathbb{Z}} \exp(-2\pi^2 m j^2 - i2\pi m j \log t) \right) \\
&= \frac{\log 2\pi}{2m} + \frac{\log m}{2m} + \frac{1}{2} \log^2 t \\
&\quad + \frac{1}{m} \log\left(\sum_{j \in \mathbb{Z}} \exp(-2\pi^2 m j^2 - i2\pi m j \log t) \right) \\
&= \frac{1}{2} \log^2 t + r_m + \frac{1}{m} \log\left(1 + \widetilde{R_m}(t)\right),
\end{aligned}
\tag{17}
$$

with

$$
r_m = \frac{\log 2\pi}{2m} + \frac{\log m}{2m},
\tag{18}
$$

$$
\widetilde{R_m}(t) = 2 \sum_{j=1}^{\infty} \exp(-2\pi^2 m j^2) \cos(2\pi m j \log t).
\tag{19}
$$

Thus

$$
\begin{aligned}
\lim_{m \to \infty} f_m(t) &= \lim_{m \to \infty} \frac{1}{m} \log \sum_{j \in \mathbb{Z}} \hat{f}_m(j;t) \\
&= \frac{1}{2} \log^2 t + \lim_{m \to \infty} \frac{1}{m} \log(1 + 2\widetilde{R_m}(t)).
\end{aligned}
\tag{20}
$$

Next, we observe that (19) implies that for every $k \in \mathbb{Z}_+$ one can find $C_k > 0$ such that $\widetilde{R_m}^{(k)}(t) := (\frac{d}{dt})^k \widetilde{R_m}(t)$ satisfies the following estimate

$$
\left| \widetilde{R_m}^{(k)}(t) \right| \leq C_k t^{-k} \sum_{j=1}^{\infty} \exp(-2\pi^2 m j^2) m^k j^k,
$$

for $m \geq 1$, $t > 0$. Therefore, taking into account the inequalities

$$
\exp(-2\pi^2 m j^2) \leq \exp(-\pi^2 m) \exp(-\pi^2 m j^2), \quad j \geq 1,
$$
$$
z^k e^{-z} \leq k!, \quad z \geq 0,
$$

we get

$$
\begin{aligned}
\left| 2\widetilde{R_m}^{(k)}(t) \right| &\leq 2 \frac{C_k}{t^k} \exp(-\pi^2 m) m^k \sum_{j=1}^{\infty} \exp(-\pi^2 m j^2) j^k \\
&\leq 2 \frac{C_k k!}{(\pi^2 t)^k} \int_0^{+\infty} e^{-\pi^2 m \xi^2} \xi^k d\xi \\
&= 2\gamma_k \frac{C_k k!}{(\pi^2 t)^k} \frac{1}{\pi^{k+1} m^{(k+1)/2}},
\end{aligned}
\tag{21}
$$

where
$$\gamma_k := \int_0^{+\infty} e^{-\xi^2} \xi^k d\xi.$$
This concludes, in view of (17), (18), (19), the proof of our theorem. □

Let $\frac{g_m}{m}$ be the sequence of Bergmann metrics whose associated Kähler form are given by

$$\frac{\omega_m}{m} = \frac{i}{2m} \partial\overline{\partial}\left(\log(\sum_{j\in\mathbb{Z}} \exp(-\frac{j^2}{2m})|z|^{2j}) \right) dz \wedge d\overline{z}, \qquad m \in \mathbb{N}. \quad (22)$$

We propose generalization in the space of the real–analytic functions of the convergence result in the C^∞ category of Theorem 2.1 above.

Theorem 2.2. *Let C^* be endowed with the complete metric g^*. Then the sequence of Bergmann metrics $\{\frac{g_m}{m}\}_{m=1}^\infty$ satisfies the following estimates: there exist $C > 0$, $c > 0$ such that*

$$\frac{g_m}{m} - g^* = \frac{i}{2m} \partial\overline{\partial}\big(R_m(|z|^2)\big)dz \wedge d\overline{z}, \qquad m \in \mathbb{N}, \quad (23)$$

satisfies

$$\big| \partial^{\alpha_1+1}\overline{\partial}^{\alpha_2+1} \left(R_m(|z|^2) \right) \big| \leq C^{|\alpha|+1}\alpha! \frac{1}{|z|^\alpha} e^{-cm},$$

for all $\alpha = (\alpha_1, \alpha_2) \in \mathbb{Z}_+^2$, $m \in \mathbb{N}$, $z \in \mathbb{C}^*$. In particular, we derive the convergence in the space of the real analytic functions to the metric g^*.

Proof. Setting $t = |z| = \sqrt{x^2 + y^2}$, in view of (22) and the invariance under composition with the real analytic maps of the real analytic functions, it is enough to study the convergence of the sequence $f_m(t)$, $m \in \mathbb{N}$ defined by (12).

We need to derive analytic–Gevrey type estimates on $f_m(t)$ in order to show the uniform convergence in the analytic category of $f_m(t)$ for $m \to \infty$, namely, to prove more precise estimates for the term $\widetilde{R_m}(t)$. We have, by Faà di Bruno type formulas (cf. [41, 42]) that

$$\left(\frac{d}{dt}\right)^\alpha (\cos(\lambda t))$$

$$= \sum_{\ell=1}^\alpha \frac{\cos(\lambda \log t + \ell\pi/2)}{\ell!} \sum_{\substack{\alpha_1+\dots\alpha_\ell=\alpha \\ \alpha_1\geq 1,\dots,\alpha_\ell\geq 1}} \frac{\alpha!}{\alpha_1!\dots\alpha_\ell!} \prod_{q=1}^\ell \left(\frac{d}{dt}\right)^{\alpha_q} (\log(\lambda t))$$

$$= \frac{\alpha!\lambda^\alpha}{t^\alpha} \sum_{\ell=1}^\alpha (-1)^{\alpha-\ell} \frac{\cos(\lambda \log t + \ell\pi/2)}{\ell!} \Theta^{\alpha_1,\dots,\alpha_\ell}, \quad (24)$$

where

$$\Theta^{\alpha_1,\ldots,\alpha_\ell} := \sum_{\substack{\alpha_1+\alpha_\ell=\alpha \\ \alpha_1\geq 1,\ldots,\alpha_\ell\geq 1}} \frac{1}{\alpha_1\ldots\alpha_\ell}. \tag{25}$$

Applying (24) for $\lambda = 2\pi m j$, $j \in \mathbb{N}$, we get

$$\left(\frac{d}{dt}\right)^\alpha(\widetilde{R_m}(t)) = 2\sum_{j=1}^\infty \exp(-2\pi^2 m j^2)\left(\frac{d}{dt}\right)^\alpha\left(\cos(2\pi m j \log t)\right)$$

$$= 2\frac{\alpha!}{t^\alpha}\sum_{j=1}^\infty (2\pi^2 m j)^\alpha \exp(-2\pi^2 m j^2) \sum_{\ell=1}^\alpha (2\pi m j)^\ell$$

$$\times (-1)^{\alpha-\ell}\frac{\cos(2\pi m j \log t + \ell\pi/2)}{\ell!}\,\Theta^{\alpha_1,\ldots,\alpha_\ell}. \tag{26}$$

We conclude by applying analytic Gevrey estimates as in [41, 42, 43] for the expression above, using the fact that all derivatives of $\cos r$ and $\sin r$ are bounded by 1 for $r \in \mathbb{R}$. $\qquad\square$

3. Representation of Kempf's distortion function for the Kepler manifold

The (regularized) Kepler manifold [36] is (may be identified with) the $2n$-dimensional symplectic manifold (X, ω), where $X = T^*S^n \setminus 0$ the cotangent bundle to the n-dimensional sphere minus its zero section endowed with the standard symplectic form ω. This may further be identified with

$$X = \left\{(e, x) \in \mathbb{R}^{n+1} \times \mathbb{R}^{n+1} \mid e \cdot e = 1, \ x \cdot e = 0, \ x \neq 0\right\},$$

where the dot denotes the standard scalar product on $\mathbb{R}^{n+1}$. In [36] J. Souriau showed that the Kepler manifold admits a natural complex structure. Indeed he proved that by introducing

$$z = |x|e + ix \in \mathbb{C}^{n+1} = |x|(e + is), \qquad s = \frac{x}{|x|} \in S^n,$$

then X is diffeomorphic to the isotropic cone

$$C = \left\{z \in \mathbb{C}^{n+1} \mid z \cdot z = z_1^2 + \cdots + z_{n+1}^2 = 0, \ z \neq 0\right\} \subset \mathbb{C}^{n+1}$$

and hence X inherits the complex structure of C via this diffeomorphism. Seven years later J. Rawnsley [15] observed that the symplectic form ω is indeed a Kähler form with respect to this complex structure and it can be written (up to a factor) as

$$\omega = \frac{i}{2}\partial\bar\partial|x|. \tag{27}$$

Moreover, since ω is exact, it is trivially integral and hence there exists a holomorphic line bundle L over X such that $c_1(L) = [\omega]$.

For $n \geq 3$, X is simply-connected so L^m is holomorphically trivial ($L^m = X \times \mathbb{C}$) and we can identify $H^0(L^m)$ with the set of holomorphic functions of X. Furthermore, we can define an Hermitian metric h_m on $L^m = X \times \mathbb{C}$ by

$$h_m(\sigma(z), \sigma(z)) = e^{-m|x|}, \qquad (28)$$

where $\sigma : X \to X \times \mathbb{C}$, is the global holomorphic section such that $\sigma(z) = (z, 1)$. It follows by (1) above that the pair (L^m, h_m) is indeed a geometric quantization of the Kepler manifold (X, ω). Then the Hilbert space $\mathcal{H}_m$ consists of the set of holomorphic functions f of X such that

$$\|f\|_m^2 := \int_X |f(z)|^2 e^{-m|x|} d\mu(z) < \infty,$$

where

$$d\mu(z) = \frac{\omega^n(z)}{n!} = \left(\frac{i}{2}\partial\bar{\partial}|x|\right)^n.$$

Notice that, in this case,

$$T_m(z) = e^{-m|x|} K^{(m)}(z, z),$$

where $K^{(m)}(z, z)$ is the reproducing kernel for the Hilbert space $\mathcal{H}_m$. At p. 412 in [15] Rawnsley explicitly computed $K(z, z) = K^{(1)}(z, z)$ (the reproducing kernel for $\mathcal{H} = \mathcal{H}_1$) and hence the corresponding Kempf's distortion function, which in our notations reads as:

$$T_1(z) = e^{-|x|} K(z, z) = \pi^{-n} 2^{n-1} e^{-|x|} \sum_{j=0}^{\infty} \frac{(j + n - 2)!}{(2j + n - 2)!} \frac{|x|^{2j}}{j!}, \qquad (29)$$

with $2|x|^2 = z \cdot \bar{z}$. Now, we compute the Kempf distortion functions $T_m(z) = T_m(|x|)$ for all non-negative integers m as follows. Making the change of variable $mz = w$, we get

$$\|f\|_m^2 = \int_X |f(w/m)|^2 e^{-|\operatorname{Im} w|} m^{-n} d\mu(w),$$

since $d\mu(w/m) = m^{-n} d\mu(w)$. Consequently, the operator

$$T : Tf(w) := m^{-\frac{n}{2}} f(w/m)$$

is a unitary isomorphism from $\mathcal{H}_m$ onto $\mathcal{H}$. Denoting by $K^{(m)}(w, z) \equiv K_z^m(w)$ the reproducing kernel of $\mathcal{H}_m$ (and writing simply $K(w, z) \equiv K_z(w)$ if $m = 1$), we therefore have, on the one hand,

$$f(z) = \langle f, K_z^{(m)} \rangle_m = \langle Tf, TK_z^{(m)} \rangle$$

for any $f \in \mathcal{H}_m$, while, on the other hand,

$$f(z) = m^{\frac{n}{2}} T f(mz) = \langle Tf, m^{\frac{n}{2}} \mathrm{K}_{mz} \rangle.$$

Thus $T \mathrm{K}_z^{(m)} = m^{\frac{n}{2}} \mathrm{K}_{mz}$, and

$$\mathrm{K}_z^{(m)}(w) = m^{\frac{n}{2}} T^{-1} \mathrm{K}_{mz}(w) = m^n \mathrm{K}_{mz}(mw).$$

That is,

$$\mathrm{K}^{(m)}(w, z) = m^n \mathrm{K}(mw, mz).$$

Substituting this into Ranwsley's formula (29), we thus get

$$\begin{aligned}
T_m(z) &= e^{-m|x|} \mathrm{K}^{(m)}(z, z) \\
&= \pi^{-n} 2^{n-1} m^n e^{-m|x|} \sum_{j=0}^{\infty} \frac{(j+n-2)!}{(2j+n-2)!} \frac{(m|x|)^{2j}}{j!}.
\end{aligned} \tag{30}$$

One observes that (30) implies that for $x = |\mathrm{Im}z|$

$$T_m(|x|) = m^n T_1(m|x|). \tag{31}$$

We propose equivalent new representations of T_m which provide useful analytic informations for the study of the asymptotic behaviour of T_m for $m \to \infty$.

Theorem 3.1. *Kempf's distortion function for the Kepler manifold can be written in the following equivalent forms:*

i) *We have*

$$T_m(|x|) = T_m^0(|x|) + \widetilde{T_m}(|x|) + \pi^{-n}(2^{n-1} - 2)m^n e^{-m|x|}, \tag{32}$$

where

$$T_m^0(|x|) = \pi^{-n} 2^n m^n e^{-m|x|} \cosh(m|x|) = \pi^{-n} m^n (1 + e^{-2m|x|}), \tag{33}$$

$$\widetilde{T_m}(|x|) = 2\pi^{-n} m^n e^{-m|x|} \sum_{j=1}^{\infty} \tau_j \frac{(m|x|)^{2j}}{(2j)!}, \tag{34}$$

$$\tau_j = \frac{(j+1) \dots (j+n-2)}{(j+1/2) \dots (j+(n-2)/2)} - 1, \qquad j \in \mathbb{N}. \tag{35}$$

ii) *Set $\xi_m = m|x|$. Then $T_m(|x|)$ equals*

$$\pi^{-n} m^n e^{-\xi_m} \left(\frac{1}{\xi_m} \frac{\partial}{\partial \xi_m} \right)^{n-2} \left(\xi_m^{n-2}(e^{\xi_m} + (-1)^n e^{-\xi_m} - 2Q(\xi_m)) \right), \tag{36}$$

where Q is polynomial of degree $[(n-4)/2]$, defined by

$$Q(\xi_m) = \left(\frac{\partial}{\partial \xi_m}\right)^{(1-(-1)^n)/2} \left(\sum_{j=0}^{\left[\frac{n-4}{2}\right]} \frac{\xi_m^{2j}}{(2j)!} \right) \qquad (37)$$

and $Q = 0$ if $n = 3$. Here $[t]$ stands for the integer part of $t \in \mathbb{R}$.

Proof. We rewrite T_m defined by (30). We have:

$$\pi^n m^{-n} e^{m|x|} T_m(|x|)$$

$$= 2^{n-1} + 2^{n-1} \sum_{j=1}^{\infty} \frac{(j+n-2)!}{(2j+n-2)!} \frac{(m|x|)^{2j}}{j!}$$

$$= 2^{n-1} + 2^{n-1} \sum_{j=1}^{\infty} \frac{(j+n-2)!(2j)!}{j!(2j+n-2)!} \frac{(m|x|)^{2j}}{(2j)!}$$

$$= 2^{n-1} + 2^{n-1} \sum_{j=1}^{\infty} \frac{(j+1)\dots(j+n-2)}{(2j+1))\dots(2j+n-2)} \frac{(m|x|)^{2j}}{(2j)!}$$

$$= 2^{n-1} + 2 \sum_{j=1}^{\infty} \frac{(j+1)\dots(j+n-2)}{(j+1/2))\dots(j+(n-2)/2)} \frac{(m|x|)^{2j}}{(2j)!}$$

$$= 2\cosh(m|x|) + (2^{n-1} - 2) + 2 \sum_{j=1}^{\infty} \tau_j \frac{(m|x|)^{2j}}{(2j)!}, \qquad (38)$$

where $\cosh t = (e^t + e^{-t})/2$ stands for the hyperbolic cosine. Clearly (38) implies the representation i).

The identity (36) has been proved in [1] with separate expressions for n even and n odd. The unified representation (37) is deduced easily, setting, as in [1]

$$y_m^2 = m|x| = \xi_m, \qquad y_m, \xi_m \in \mathbb{R} \setminus \{0\},$$

and writing

$$T_m(|x|) = 2^{n-1} \pi^{-n} m^n e^{-\xi_m} \sum_{j=0}^{\infty} \frac{(j+n-2)!}{(2j+n-2)!} \frac{y_m^j}{j!}$$

$$= 2^{n-1} \pi^{-n} m^n e^{-\xi_m} \left(\frac{\partial}{\partial y_m}\right)^{n-2} \sum_{j=0}^{\infty} \frac{y_m^{j+n-2}}{(2j+n-2)!} \qquad (39)$$

$$= 2\pi^{-n} m^n e^{-\xi_m} \left(\frac{1}{\xi_m} \frac{\partial}{\partial \xi_m}\right)^{n-2} \left(\xi_m^{n-2} \sum_{j=0}^{\infty} \frac{\xi_m^{2j+n-2}}{(2j+n-2)!} \right). \qquad (40)$$

The novelty in comparison to [1] is the observation that, setting $\varkappa_n = (1 - (-1)^n)/2$, i.e. $\varkappa_n = 0$ (respectively, $\varkappa_n = 1$) if n is even (respectively, odd), we have

$$2 \sum_{j=0}^{\infty} \frac{\xi_m^{2j+n-2}}{(2j+n-2)!} = e^{-\xi_m} + (-1)^n e^{-\xi_m}$$

$$+ 2\left(\frac{\partial}{\partial \xi_m}\right)^{(1-(-1)^n)/2} \left(\sum_{j=0}^{\left[\frac{n-4}{2}\right]} \frac{\xi_m^{2j}}{(2j)!} \right). \quad (41)$$

Clearly if $n \geq 4$ is even

$$Q(\xi_m) = \sum_{j=0}^{\frac{n-4}{2}} \frac{\xi_m^{2j}}{(2j)!},$$

while

$$R(\xi_m) = \sum_{j=0}^{\frac{n-5}{2}} \frac{\xi_m^{2j+1}}{(2j+1)!},$$

if $n \geq 5$ is odd, recapturing the expressions in [1]. □

Remark 3.1. We note that (35) implies

$$\lim_{j \to \infty} \tau_j = 0$$

which, in view of the new representation (32), allows us to obtain the principal term m^n of the TYZ expansion of $T_m(|x|)$.

4. TYZ expansion for the Kepler manifold

We derive the TYZ expansion of T_m for the Kempf distortion function of the Kepler manifold by using (36), cf. [1]. The explicit representation of $T_m(|x|)$ for the Kepler manifold has a remarkable feature, namely, it is defined by a generating function depending on one variable $\rho = |x|$. We derive the TYZ expansion for the Kepler manifold and obtain better estimates of the remainder for the asymptotic expansion of T_m in comparison with [1].

Theorem 4.1. *The following representation of T_m holds*

$$T_m(|x|) = \pi^{-n} m^n \sum_{j=0}^{n-2} \frac{a_j}{m^j |x|^j} + 2\pi^{-n} m^n E(m|x|), \quad m \in \mathbb{N}, \quad (42)$$

where

$$a_j \in \mathbb{R}, \qquad j = 0, 1, \ldots, n-2, \quad (43)$$

with

$$a_0 = 1, \qquad a_1 = \frac{(n-2)(n-1)}{2}, \tag{44}$$

$$E(y) = e^{-2y} \sum_{j=0}^{n-2} \frac{p_j}{y^j} + e^{-y} \sum_{k=0}^{n-3} \frac{r_k}{y^k}, \tag{45}$$

for some $p_j \in \mathbb{R}$, $r_k \in \mathbb{R}$, $j = 0, 1, \ldots, n-2$, $k = 0, 1, \ldots, n-3$. Moreover, there exists an absolute constant $C_0 > 0$ such that the remainder

$$m^n E(m\rho) = T_m(\rho) - \sum_{j=0}^{n-2} \frac{a_j}{m^j \rho^j}$$

satisfies the following global analytic–Gevrey estimates

$$|D_\rho^\alpha(m^n E(m\rho))| \le C_0^{\alpha+1} \alpha! (1 + \frac{1}{\rho^{n-2+\alpha}}) m^n \exp(-\frac{1}{2} m\rho), \tag{46}$$

for all $m \in \mathbb{N}$, $\rho = |x| > 0$, $\alpha \in \mathbb{Z}_+$.

Proof. We derive (46). The other assertions are proved in [1]. Let $k \in \mathbb{N}$, $c > 0$. Applying the Leibniz formula, we get the following by setting $D_\rho = \frac{d}{d\rho}$:

$$\begin{aligned}
S_c^\alpha(\rho) &:= D_\rho^\alpha \left(\rho^{-\alpha} \exp(-c\rho) \right) \\
&= \sum_{\beta=1}^\alpha \frac{\alpha!}{\beta!(\alpha-\beta)!} D_\rho^\beta(\rho^{-k})(-c)^{\alpha-\beta} \exp(-c\rho) \\
&= (-1)^\alpha \alpha! \sum_{\beta=1}^\alpha \frac{k \ldots (k+\beta-1)}{\beta!(\alpha-\beta)!} \rho^{-k-\beta} c^{\alpha-\beta} \exp(-c\rho) \\
&= \frac{\alpha!}{t^{k+\alpha}} \sum_{\beta=1}^\alpha \frac{(-1)^\beta k \ldots (k+\beta-1)}{\beta!} \frac{\rho^{\alpha-\beta}}{(\alpha-\beta)!} \exp(-c\rho). \tag{47}
\end{aligned}$$

Combining (47) with combinatorial inequalities (e.g., see [41, 42, 43]) we get that

$$\begin{aligned}
S_c^\alpha(\rho) &\le \sum_{\beta=1}^\alpha \frac{\alpha!}{\beta!(\alpha-\beta)!} \left| D_\rho^\beta(\rho^{-k})(-c)^{\alpha-\beta} \right| \exp(-c\rho) \\
&= \frac{\alpha!}{\rho^{k+\alpha}} \exp(-\frac{1}{2}c\rho) \sum_{\beta=1}^\alpha \frac{k \ldots (k+\beta-1)}{\beta!} \left(\frac{(c\rho)^{\alpha-\beta}}{(\alpha-\beta)!} \exp(-\frac{1}{2}c\rho) \right) \\
&\le 2^{k+\alpha} \frac{\alpha!}{\rho^{k+\alpha}} \exp(-\frac{1}{2}c\rho), \tag{48}
\end{aligned}$$

for all $\rho > 0$, $c > 0$, $\alpha \in \mathbb{Z}_+$. We conclude the proof by applying the estimate (48) for each term of $E(m\rho)$ with $k = 1, \ldots, n - 2$ and taking into account that

$$\max\left\{1, \frac{1}{\rho}, \ldots, \frac{1}{\rho^{n-2}}\right\} \le 1 + \frac{1}{\rho^{n-2}}, \qquad \rho > 0.$$

The constant C_0 depends on $|p_j|$, $j = 0, 1, \ldots, n-2$, $|r_k|$, $k = 0, 1, \ldots, n-3$ and n. $\qquad\square$

Remark 4.1. We observe that the global analytic–Gevrey estimates imply the corresponding estimates on the remainder in [1]. Next, we point out that another interesting problem concerns the asymptotic expansion and the estimates of the "obstruction" $\mathcal{E}_m(x)$ in (8) for the approximation of ω (resp. g) by $\frac{\varphi_m^*(\omega_{FS})}{m}$ (respectively, $\frac{\varphi_m^*(g_{FS})}{m}$) as in (7). We address this issue in our work [44].

Acknowledgments

The authors thank the unknown referee for the critical remarks which led to improvements in the final form of the paper.

The first author was supported in part by GNAMPA-INDAM, Italy and the M.I.U.R. Project "Microlocal Analysis".

The second author was supported in part by the M.I.U.R. Project "Geometric Properties of Real and Complex Manifolds"

References

1. T. Gramchev, A. Loi, *TYZ expansion for the Kepler manifold*, Comm. Math. Phys. **289** (2009), 825-840.
2. C. Arezzo and A. Loi, *Quantization of Kähler manifolds and the asymptotic expansion of Tian–Yau–Zelditch*, J. Geom. Phys. **47** (2003), 87-99.
3. C. Arezzo and A. Loi, *Moment maps, scalar curvature and quantization of Kähler manifolds*, Comm. Math. Phys. **246** (2004), 543-549.
4. S. Ji, *Inequality for distortion function of invertible sheaves on Abelian varieties*, Duke Math. J. **58** (1989), 657-667.
5. G. R. Kempf, *Metric on invertible sheaves on abelian varieties*, Topics in algebraic geometry (Guanajuato) (1989).
6. A. Loi, D. Zuddas, *Some remarks on Bergmann metrics*, Riv. Mat. Univ. Parma **6**, no. 4 (2001), 71-86.
7. A. Loi, *The Tian–Yau–Zelditch asymptotic expansion for real analytic Kähler metrics*, Int. J. of Geom. Methods Mod. Phys. **1** (2004), 253-263.
8. A. Loi, *A Laplace integral, the T-Y-Z expansion and Berezin's transform on a Kaehler manifold*, Int. J. of Geom. Methods Mod. Phys. **2** (2005), 359-371.

9. A. Loi, *Regular quantizations of Kaehler manifolds and constant scalar curvature metrics*, J. Geom. Phys. **53** (2005), 354-364.

10. A. Loi, *Bergman and balanced metrics on complex manifolds*, Int. J. Geom. Methods Mod. Phys. **2** (2005), 553-561.

11. A. Loi, *Calabi's diastasis function for Hermitian symmetric spaces*, Differential Geom. Appl. **24** (2006), 311-319.

12. A. Loi, F. Cuccu, *Balanced metrics on $\mathbb{C}^n$*, J. Geom. Phys. **57** (2007), 1115-1123.

13. A. Loi, *Regular quantizations and covering maps*, Geom. Dedicata **123** (2006), 73-78.

14. J. H. Rawnsley, *A nonunitary pairing of polarizations for the Kepler problem*, Trans. Amer. Math. Soc. **250** (1979), 167-180.

15. J. H. Rawnsley, *Coherent states and Kähler manifolds*, Quart. J. Math. Oxford Ser. (2) **28**(1977), 403-415.

16. W. D. Ruan, *Canonical coordinates and Bergmann metrics*, Comm. Anal. Geom. **6** (1998), 589-631.

17. G. Tian, *On a set of polarized Kähler metrics on algebraic manifolds*, J. Differential Geom. **32** (1990), 99-130.

18. S. Zelditch, *Szegö Kernels and a Theorem of Tian*, Internat. Math. Res. Notices **6** (1998), 317-331.

19. S. Zhang, *Heights and reductions of semi-stable varieties*, Comp. Math. **104** (1996), 77-105.

20. Z. Lu, *On the lower terms of the asymptotic expansion of Tian–Yau–Zelditch*, Amer. J. Math. **122** (2000), 235-273.

21. E. Calabi, *Isometric Imbeddings of Complex Manifolds*, Ann. of Math. **58** (1953), 1-23.

22. F. A. Berezin, *Quantization*, Izv. Akad. Nauk SSSR Ser. Mat. **38** (1974), 1116–1175 (Russian).

23. F. A. Berezin, *Quantization in complex symmetric spaces*, Izv. Akad. Nauk SSSR Ser. Mat. **39** (1975), 363–402 (Russian).

24. M. Engliš, *Berezin Quantization and Reproducing Kernels on Complex Domains*, Trans. Amer. Math. Soc. **348** (1996), 411-479.

25. M. Engliš, *A Forelli–Rudin contruction and asymptotics of weighted Bergman kernels*, J. Funct. Anal. **177** (2000), 257-281.

26. M. Engliš, *The asymptotics of a Laplace integral on a Kähler manifold*, J. Reine Angew. Math. **528** (2000), 1-39.

27. M. Cahen, S. Gutt and J. H. Rawnsley, *Quantization of Kähler manifolds I: Geometric interpretation of Berezin's quantization*, J. Geom. Phys. **7** (1990), 45-62.

28. M. Cahen, S. Gutt and J. H. Rawnsley, *Quantization of Kähler manifolds II*, Trans. Amer. Math. Soc. **337** (1993), 73-98.

29. M. Cahen, S. Gutt and J. H. Rawnsley, *Quantization of Kähler manifolds III*, Lett. Math. Phys. **30** (1994), 291-305.

30. M. Cahen, S. Gutt and J. H. Rawnsley, *Quantization of Kaehler manifolds IV*, Lett. Math. Phys. **34** (1995), 159-168.

31. C. Moreno and P. Ortega-Navarro, *∗-products on $D^1(C)$, S^2 and related*

spectral analysis, Lett. Math. Phys. **7** (1983), 181-193.

32. C. Moreno, *Star-products on some Kähler manifolds*, Lett. Math. Phys. **11** (1986), 361-372.

33. A. Odzijewicz, *On reproducing kernels and quantization of states*, Commun. Math. Phys. **114** (1988), 577-597.

34. A. Odzijewicz, *Coherent states and geometric quantization*, Commun. Math. Phys. **150** (1992), 385-413.

35. I. M. Mladenov and V. V. Tsanov, *Reduction in stages and complete quantization of the MIC-Kepler problem* J. Phys. A **32** (1999), 3779–3791.

36. J. M. Souriau, *Sur la varie'te' de Kepler*, Symposia Mathematica XIV, (Academic Press, 1974).

37. S. Donaldson, *Scalar Curvature and Projective Embeddings, I*, J. Differential Geom. **59** (2001), 479-522.

38. S. Donaldson, *Scalar Curvature and Projective Embeddings, II*, Q. J. Math. **56** (2005), 345–356.

39. S. Donaldson, *Some numerical results in complex differential geometry*, Pure Appl. Math. Q. (Special Issue: In honor of Friedrich Hirzebruch. Part 1) **5** (2009), 571-618.

40. R. Godement, *Analyse Mathématique II*, (Springer, 1998).

41. M. Cappiello, T. Gramchev and L. Rodino, *Super-exponential decay and holomorphic extensions for semilinear equations with polynomial coefficients*, J. Func. Anal. **237** (2006), 634-654.

42. M. Cappiello, T. Gramchev and L. Rodino,*Entire extensions and exponential decay for semilinear elliptic equations*, Journal d'Analyse Mathematique **111** (2010), 339-367.

43. F. Cardin, T. Gramchev and A. Lovison, *Exponential estimates for oscillatory integrals with degenerate phase functions*, Nonlinearity, **21** (2008), 409-433 .

44. T. Gramchev and A. Loi, *On the remainder term of the TYZ expansion for the Kepler manifold*, preprint (2011).

Received January 31, 2011
Revised   April 30, 2011

Proceedings of the 2nd International
Colloquium on Differential Geometry
and its Related Fields
Veliko Tarnovo, September 6–10, 2010

KERSHNER'S TILINGS OF TYPE 6 BY CONGRUENT PENTAGONS ARE NOT DIRICHLET

Atsushi KUBOTA

*Department of Computer Science, Nagoya Institute of Technology,
Nagoya 466-8555, Japan*

Toshiaki ADACHI*

*Department of Mathematics, Nagoya Institute of Technology,
Nagoya 466-8555, Japan
E-mail: adachi@nitech.ac.jp*

In this note we show that tessellations of tiling-type 6 of a plane by congruent pentagons are not Dirichlet.

Keywords: Convex tessellations, Dirichlet tessellations, Kershner's pentagons, Tiling-type, Cites.

1. Introduction

A family $\mathcal{R} = \{R_\alpha\}$ of convex closed polygons in a Euclidean plane $\mathbb{R}^2$ is said to be a convex tessellation (or a convex tiling) if it covers this plane (i.e. $\mathbb{R}^2 = \bigcup_\alpha R_\alpha$) and satisfies the following condition: If $R_\alpha \cap R_\beta \neq \emptyset$, then this set $R_\alpha \cap R_\beta$ is either an edge of each R_α and R_β or a vertex of each of them. This condition means that two polygons might meet by their boundary and that every vertex of polygons is not an intermediate point of an edge of any other polygon. One of most simplest ways to get a convex tessellation is to make a Voronoi diagram: Take a set $\mathcal{P} = \{P_\alpha\}$ of points on $\mathbb{R}^2$ and define each region R_α by

$$R_\alpha = \bigcap_{\beta \neq \alpha} \{P \in \mathbb{R}^2 \mid d(P, P_\alpha) \leq d(P, P_\beta)\}.$$

The set $\mathcal{P}$ is called the set of *cites* for this Voronoi diagram. Such tessellations are used and studied in the areas of information geometry, biology, economics and some other fields of science.

*The second author is partially supported by Grant-in-Aid for Scientific Research (C) (No. 20540071) Japan Society of Promotion Science.

A convex tessellation $\mathcal{R}$ of $\mathbb{R}^2$ is said to be *Dirichlet* if there is a set of points $\mathcal{P}$ in $\mathbb{R}^2$ satisfying that the tessellation of the Volonoi diagram determined by this set $\mathcal{P}$ coincides with this tessellation. Therefore, for a Dirichlet tessellation, the choice of its cites may not be unique. It is an interesting problem to get a condition for tessellations to be Dirichlet. Given a convex tessellation $\mathcal{R}$ we say a vertex of a polygon to be n-valent if n-polygons meet at that vertex, in another words, if n-edges of the tessellation emanate from this vertex. In their paper [1] Ash and Bolker gave a necessary and sufficient condition for a tessellation all of whose vertices are of odd-valent to be Dirichlet. But the situation is not so simple for tessellations containing even-valent edges.

Since our problem is complicated for general tessellations, we restrict ourselves to tessellations obtained by polygons which are congruent to each other. In [5] Takeo and the second author studied the Dirichlet property for some patterns of tessellations by congruent quadrangles. In order to proceed our study we need to take tessellations by congruent polygons. In 1968, Kershner [2] showed three new patterns of paving a plane by congruent convex pentagons. In this note we take a pattern of tiling which is called of type 6 (see §2), and show that it is not Dirichlet. We also show that tessellations of tiling-type 6 which is closely related with Kershner's tilings do not satisfy the Dirichlet condition.

The authors are grateful to Mr. Y. Takeo for his assistance in drawing Kershner's tilings.

2. Kershner's tilings of type 6

We here recall tessellations of type 6 by congruent convex pentagons which were given by Kershner. Let $R = \text{ABCDE}$ be a convex pentagon whose angles at vertices and whose lengths of edges satisfy

$$\angle A + \angle B + \angle D = 2\pi, \quad \angle A = 2\angle C$$

and

$$\overline{AB} = \overline{DE} = \overline{EA}, \quad \overline{BC} = \overline{CD}$$

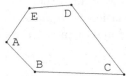

Fig. 1. Kershner's pentagon

(see Fig. 1). In order to pave a plane we consider a fundamental region consists by four such pentagons. We take four pentagons R_i = $A_iB_iC_iD_iE_i$ ($i = 1, 2, 3, 4$) which are congruent to R. Like Fig. 2, we paste R_1 and R_2 by their edges as $B_1C_1 = D_2C_2$, R_2 and R_3 by their edges as $B_2C_2 = C_3B_3$, R_3 and R_4 by their edges as $B_3A_3 = D_4E_4$, and R_4 and R_1 by their edges as $C_4D_4 = D_1C_1$. We then obtain a fundamental region $\mathcal{F} = R_1 \cup R_2 \cup R_3 \cup R_4$. One can easily see that a family of such fundamental regions pave a plane (see Fig. 3). We here note that a Kershner's pentagon is not regular. Since angles of regular pentagons are $3\pi/5$, they cannot make the angle 2π, hence cannot pave a plane.

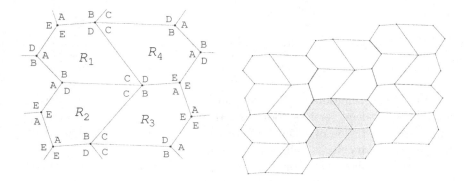

Fig. 2.   Fundamental domain   Fig. 3.   Kershner's tiling of type 6

3. The Dirichlet property of Kershner's tilings of type 6

In this section we show the following result.

Theorem 3.1. *Kershner's tilings of a plane* $\mathbb{R}^2$ *of type* 6 *are not Dirichlet.*

By Fig. 2, we find each pentagon in a Kershner's tiling of type 6 has three 3-valent vertices. These vertices are formed by either vertices (A, B, D) or vertices (A, E, E). Every 4-valent vertex is formed by vertices (B, C, C, D). These pentagons in a Kershner's tiling of type 6 are divided into two classes;

  i) pentagons having two 3-valent vertices of type (A, B, D) and one 3-valent vertex of type (A, E, E),
  ii) pentagons having two 3-valent vertices of type (A, E, E) and one 3-valent vertex of type (A, B, D).

We call the former pentagons of adjacent type (I) and the latter pentagons of adjacent type (II).

If a tessellation is Dirichlet and has an odd-valent vertices, we have information on angles formed by cites and odd-valent vertices. Here, we write down the situation around a 3-valent vertex (see [1, 6]).

Lemma 3.1. *There are three vertices* P_0, P_1, P_2
and three edges OA_0, OA_1 *and* OA_2 *which meet at*
the vertex O *on* $\mathbb{R}^2$. *If we suppose the line* OA_i *is*
the perpendicular bisector of the segment $P_{i-1}P_i$
for each i, *where indices are considered in modulo*
3, then we have $\overline{OP_0} = \overline{OP_1} = \overline{OP_2}$ *and*

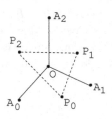

$$\angle P_i OA_i = \frac{1}{2}(\theta_i - \theta_{i+1} + \theta_{i+2}) = \angle P_{i+2} OA_i \quad (1)$$

where $\theta_i = \angle A_i OA_{i+1}$.

Fig. 4.   Positions of cites
around 3-valent vertex

When a tessellation is Dirichlet, alternating sums of angles like (1) around odd valent vertices show directions of cites from those vertices. Therefore, if a region of a Dirichlet tessellation has two odd vertices, then the position of its cite is uniquely determined.

In order to show Theorem 3.1, we suppose a Kershner's tiling of type 6 to be Dirichlet and point out a contradiction. Given a Kershner's pentagon $R = ABCDE$ we take points $P, Q \in R$ satisfying

$$\angle PED = \angle C, \quad \angle PBC = (\angle B - \angle A + \angle D)/2, \quad \angle QAE = \angle QEA = \angle C.$$

We say two pairs (R_1, P_1), (R_2, P_2) of pentagons and points are congruent to each other if there is a congruent motion $f : \mathbb{R}^2 \to \mathbb{R}^2$ satisfying $f(R_1) = R_2$ and $f(P_1) = P_2$.

We suppose our Kershner's tiling $\mathcal{R}$ of type 6 is Dirichlet. Since $\angle A = 2\angle C$, when a pentagon $R_1 = A_1 B_1 C_1 D_1 E_1$ is of adjacent type (I), its cite $P_1 \in R_1$ satisfies

$$\angle P_1 B_1 C_1 = (\angle B - \angle A + \angle D)/2,$$
$$\angle P_1 E_1 D_1 = (\angle E - \angle E + \angle A)/2 = \angle C$$

by Lemma 3.1 (see Fig. 5). Similarly, when a pentagon $R_2 = A_2 B_2 C_2 D_2 E_2$

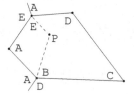

Fig. 5.   A pentagon of adjacent type (I)

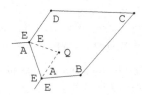

Fig. 6.   A pentagon of adjacent types (II)

is of adjacent type (II), its cite $Q_2 \in R_2$ satisfies

$$\angle Q_2 A_2 E_2 = (\angle A - \angle E + \angle E)/2 = \angle C,$$
$$\angle Q_2 E_2 A_2 = (\angle E - \angle E + \angle A)/2 = \angle C$$

(see Fig. 6). Therefore we find that pentagons of adjacent type (I) with their cites are congruent to $(ABCDE, P)$ and that pentagons of adjacent type (II) with their cites are congruent to $(ABCDE, Q)$.

We now study at 4-valent vertices. We take a fundamental domain $\mathcal{F}$ like Fig. 2. In that figure we see R_1, R_4 are of type (I) and R_2, R_3 are of type (II). We denote their cites by P_1, P_4, Q_2 and Q_3. Since the edge $B_1 C_1 = D_2 C_2$ is the perpendicular bisector of the edge $P_1 Q_2$, we find $\angle P_1 C_1 B_1 = \angle Q_2 C_2 D_2$. Applying the same argument to pairs of pentagons $R_2 : R_3$, $R_3 : R_4$, $R_4 : R_1$, we obtain

$$\angle Q_2 C_2 B_2 = \angle Q_3 B_3 C_3, \ \angle Q_3 B_3 A_3 = \angle P_4 D_4 E_4, \ \angle P_4 D_4 C_4 = \angle P_1 C_1 D_1.$$

As we have

$$\angle P_1 C_1 B_1 + \angle P_1 C_1 D_1 = \angle C = \angle Q_2 C_2 D_2 + \angle Q_2 C_2 B_2,$$

we find $\angle Q_3 B_3 C_3 = \angle P_4 D_4 C_4$, hence obtain

$$\angle B = \angle Q_3 B_3 C_3 + \angle Q_3 B_3 A_3 = \angle P_4 D_4 C_4 + \angle P_4 D_4 E_4 = \angle D.$$

Coming back to the original pentagon, we find two triangles $\triangle ABC$, $\triangle EDC$ are congruent to each other, because $\overline{AB} = \overline{ED}$, $\overline{BC} = \overline{DC}$. We therefore obtain $\angle A = \angle E$. By use of original conditions on angles of the original pentagon, we can conclude that $\angle A = \angle B = \angle D = \angle E = 2\angle C = 2\pi/3$ and $P = Q$. Thus we find $\angle QCB = \angle QBC = \angle PBC = \pi/6 = \angle C$. This means that $P = Q$ lies on the edge CD, which is a contradiction. We hence find that Kershner's tilings of type 6 are not Dirichlet.
This complete the proof of Theorem 3.1.

Remark 3.1. By use of a pentagon $R = ABCDE$ with $\angle A = \angle B = \angle D = \angle E = 2\angle C = 2\pi/3$, we can make another tessellation which is called of type 5 like Fig. 7. This tessellation has 3-valent vertices and 6-valent vertices. By taking a cite of each pentagon $ABCDE$ at the mid point of the segment BD, we find it is a Dirichlet tessellation.

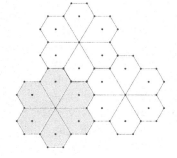

Fig. 7.   A tessellation of type 5

4. Tessellations of type 6 by congruent pentagons

When we consider Kershner's tilings, they are formed by only one side (i.e. front face) of congruent tiles of pentagons. But when we consider tessellations of a plane by congruent polygons, we allow to use both the front face and the back face. It was pointed out by Sugimoto and Ogawa [4] that there are infinitely many patterns of tessellations which were paved by congruent Kershner's pentagons. Such tessellations of a plane by pentagons all of which are congruent to a Kershner's pentagon are called of type 6 (see [3]).

When we consider two faces of a Kershner's pentagon $R = $ ABCDE, we may suppose $\angle B \neq \angle D$. Since it satisfies $\overline{AB} = \overline{DE} = \overline{EA}$, $\overline{BC} = \overline{CD}$, in the case $\angle B = \angle D$, by the same argument as in the previous section, we get $\angle A = \angle E$ and find that it is the pentagon in Remark 3.1. In this case we need not to consider its two faces.

In order to pave a plane by pentagons congruent to a Kershner's pentagon R, as it is not a regular pentagon, the conditions on lengths of edges show that an edge BC meets either an edge BC or an edge CD of another pentagon. Hence a vertex C do not meet with vertices A and E of other pentagons. As R also satisfies $\angle A + \angle B + \angle D = 2\pi$, $\angle A + 2\angle E = 2\pi$ (hence $\angle A = 2\angle C$), without any additional conditions, we see that every tessellation by pentagons congruent to R has a 4-valent vertex formed by $\angle B$, $\angle D$ and two $\angle C$ and that at that vertex $\angle B$ and $\angle D$ are faced and two $\angle C$ are faced. Therefore, if four pentagons meet and make a 4-valent vertex, the situation around this vertex is either like the pattern of Kershner's fundamental domain $\mathcal{F}$ or like the pattern $\mathcal{G}$

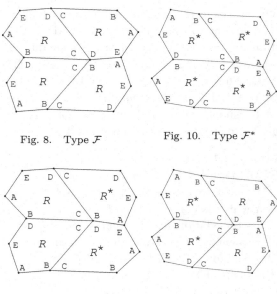

Fig. 8.   Type $\mathcal{F}$

Fig. 10.   Type $\mathcal{F}^*$

Fig. 9.   Type $\mathcal{G}$

Fig. 11.   Type $\mathcal{G}^*$

(Fig. 9). In those Figs. 9, 10, 11, we denote a reversed pentagon by R^*. Of course, we regard two patterns are equivalent to each other if they are congruent to each other under actions of rotations and reflections. The domain $\mathcal{F}^*$ coincides with $\mathcal{F}$ by a reflection, so does $\mathcal{G}^*$ with $\mathcal{G}$. Considering 4-valent vertices we find the upside and the downside of a fundamental domain should be of the same type to pave a plane. Hence, we only need to consider patterns how to past pentagons to the left-side and to the right-side of these patterns of fundamental domains made by four pentagons.

Next we consider at vertices A and E. We study the way to put pentagons to the right-hand side of the domains $\mathcal{F}$ and $\mathcal{G}$. By the conditions on a Kershner's pentagon we obtain that we have only two patterns. If we past the left two pentagons of $\mathcal{F}^*$ to the right two pentagons of $\mathcal{F}$ without a gap, we find $\angle B + \angle D + \angle E = 2\pi$ and $2\angle A + \angle E = 2\pi$. Similarly, if we past the right two pentagons of $\mathcal{F}^*$ to the left two pentagons of $\mathcal{G}$ without a gap, we obtain these equalities. By use of the original condition on Kershner's pentagon, with one of these equality we get $\angle A = \angle E = 2\pi/3$. This guarantees that the pentagon is the one in Remark 3.1. Since we cannot recognize the difference between the front-face and the back-face of this pentagon, we only need to consider the two patterns in Fig. 12.

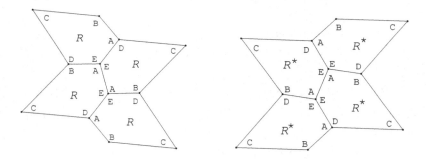

Fig. 12. Patterns of prolongations

By these considerations we can conclude that if a tessellation of type 6 has both front-faced pentagons and back-faced pentagons it contains either the pattern of eight pentagons in Fig. 13 or its π-rotation.

We are now in the position to study the Diridhlet property of tessellations of type 6.

Theorem 4.1. *Tessellations of a plane* $\mathbb{R}^2$ *of type 6 by congruent pentagons are not Dirichlet.*

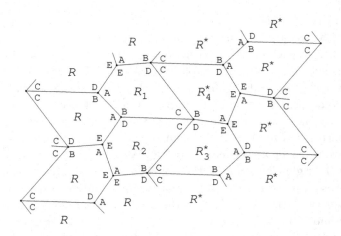

Fig. 13.   Essential part of a tessellation of type 6

Proof. By Theorem 3.1 we are enough to consider the pattern of eight pentagons in Fig. 13. We see R_1, R_3^* are of type (I) and R_2, R_4^* are of type (II). We suppose our tessellation is Dirichlet. We denote the cites of R_1, R_3^*, R_2, R_4^* by P_1, P_3^*, Q_2, Q_4^*, respectively. We denote R_1 and R_4^* by $A_1B_1C_1D_1E_1$ and $A_4^*B_4^*C_4^*D_4^*E_4^*$, respectively. Since the edge $C_1D_1 = B_4^*C_4^*$ is the perpendicular bisector of the edge $P_1Q_4^*$, we find $\angle P_1C_1D_1 = \angle Q_4^*B_4^*C_4^*$. Applying the same argument to pairs of pentagons $R_1 : R_2$, $R_2 : R_3^*$, $R_3^* : R_4^*$, we find

$$\angle P_1C_1B_1 = \angle Q_2C_2D_2, \ \angle Q_2C_2B_2 = \angle P_3^*D_3^*C_3^*, \ \angle P_3^*D_3^*E_3^* = \angle Q_4^*B_4^*A_4^*.$$

Here, we denote R_2 and R_3^* by $A_2B_2C_2D_2E_2$ and $A_3^*B_3^*C_3^*D_3^*E_3^*$, respectively. As we have

$$\angle P_1C_1B_1 + \angle P_1C_1D_1 = \angle C = \angle Q_2C_2D_2 + \angle Q_2C_2B_2,$$

we get

$$\angle B = \angle P_1C_1D_1 + \angle Q_4^*B_4^*A_4^* = \angle Q_2C_2B_2 + \angle P_3^*D_3^*E_3^* = \angle D.$$

Thus we obtain $\angle A = \angle B = \angle D = \angle E = 2\angle C = 2\pi/3$, $P = Q$ and find a contradiction by just the same way as for the proof of Theorem 3.1. We hence get the conclusion. □

References

1. P. Ash & E. D. Bolker, *Recognizing Dirichlet tessellations*, Geom. Dedicata 19(1985), 175–206.

2. R. B. Kershner, *On paving the plane*, Amer. Math. Monthly 75(1968), 839–844.

3. T. Sugimoto & T. Ogawa, *Tiling problem of convex pentagon*, Forma 15(2000), 75-79.

4. _____, *New tiling patterns of the tessellating convex pentagon (type 6)* (in Japanese), Bull. Soc. Sci. Form 15(2000), 10-21.

5. Y. Takeo & T. Adachi, *Dirichlet tessellations of a plane by congruent quadrangles*, New Zealand J. Math. 39(2009), 79–101.

6. _____, *Dirichlet property for tessellations of tiling-type 4 on a plane by congruent pentagons*, Trends in Differential Geometry and Mathematical Physics, K. Sekigawa, V. Gerdjikov and S. Dimiev eds. (2009), 219–230, World Scientific, Singapore.

Received February 18, 2011
Revised   May 3, 2011

Proceedings of the 2nd International
Colloquium on Differential Geometry
and its Related Fields
Veliko Tarnovo, September 6–10, 2010

ELEVEN CLASSES OF
ALMOST PARACONTACT MANIFOLDS WITH
SEMI-RIEMANNIAN METRIC OF $(n + 1, n)$

Galia NAKOVA

University of Veliko Turnovo "St. Cyril end St. Metodius",
Faculty of Education,
T. Tarnovski 2 str.,
5003 Veliko Tarnovo, Bulgaria
E-mail: gnakova@gmail.com

Simeon ZAMKOVOY

University of Sofia "St. Kl. Ohridski",
Faculty of Mathematics and Informatics,
Blvd. James Bourchier 5,
1164 Sofia, Bulgaria
E-mail: zamkovoy@fmi.uni-sofia.bg

In this paper we introduce eleven basic classes of almost paracontact manifolds
with semi-Riemannian metric of signature $(n + 1, n)$ and construct some examples.
MSC: 53C15, 5350, 53C25, 53C26, 53B30

Keywords: Almost paracontact manifolds, Indefinite metric.

Introduction

The notion of almost paracontact structures on differentiable manifolds
was introduced in [2]. They are analogues of almost contact structures
[3]. Almost paracontact structures and almost contact structures are odd
dimensional analogue of almost product structures and almost complex
structures, respectively. As is well known, in [5] Gray and Hervella classified tensor fields of almost Hermitian manifolds. In [1] a classification
of almost para-Hermitian manifolds is obtained. A classification of almost
paracontact Riemannian manifolds of type (n, n) is given in [8]. In this paper we consider almost paracontact manifolds with semi-Riemannian metric of signature $(n + 1, n)$, which are odd dimensional analogues of almost
para-Hermitian manifolds. Our purpose is to classify these manifolds. This

paper is organized as follows. Section 1 consists of basic definitions and notions about almost paracontact manifolds with semi-Riemannian metric of signature $(n+1, n)$. In Section 2 we give the main theorem: We obtain a classification of the considered manifolds with respect to the covariant derivatives of structure tensor fields. Sections 3, 4, 5 contain the proof of the theorem. The method used in the present paper is analogous to the one in [4]. Moreover, we characterize each of the eleven basic classes by properties of proper linear operators. Using these characterizations we construct examples of manifolds belonging to two basic classes in Section 6.

1. Preliminaries

An *almost paracontact structure* on a $(2n+1)$-dimensional smooth manifold M is a triplet (φ, ξ, η) of a tensor field φ of type $(1, 1)$, a vector field ξ and 1-form η satisfying the following conditions:

$$\varphi^2 = id - \eta \otimes \xi, \qquad \eta(\xi) = 1, \qquad \varphi(\xi) = 0, \tag{1}$$

where id denotes the identity transformation. As immediate consequences of the conditions (1) we have

$$\eta \circ \varphi = 0, \qquad \text{rank}(\varphi) = 2n.$$

Throughout this paper, by a *semi-Riemannian metric* on a manifold M we mean a non-degenerate symmetric tensor field g of type $(0, 2)$ on M. Let M be a manifold equipped with an almost paracontact structure (φ, ξ, η). Let g be a semi-Riemannian metric such that

$$g(\varphi X, \varphi Y) = -g(X, Y) + \eta(X)\eta(Y), \quad X, Y \in TM. \tag{2}$$

Such metric g is called a *compatible* metric with the given almost paracontact structure. The compatible metric g is necessarily of signature $(n+1, n)$ (see [10]). We shall say briefly that M is *an almost paracontact metric manifold* equipped with *an almost paracontact metric structure* (φ, ξ, η, g). The condition (2) is equivalent to the following conditions

$$g(X, \varphi Y) = -g(\varphi X, Y), \qquad g(X, \xi) = \eta(X), \quad X, Y \in TM.$$

For an almost paracontact metric manifold M, there exists an "orthonormal" basis $\{e_1, \varphi e_1, \ldots, e_n, \varphi e_n, \xi\}$, a basis such that

$$g(e_i, e_j) = -g(\varphi e_i, \varphi e_j) = \delta_{ij}, \quad g(e_i, \varphi e_j) = 0, \quad i, j = 1, \ldots, n,$$

(see [10]). This basis is called a φ-basis.

Let $\mathbb{U}^\pi(n)$ be the paraunitary group. That is, $\mathbb{U}^\pi(n)$ consists of para-complex matrices $\beta = A + \epsilon B$ satisfying $\beta^{-1} = \bar{\beta}^t$. Here A, B are real matrices of type $(n \times n)$, ϵ is a number with $\epsilon^2 = 1$ and $\bar{\beta}^t = A^t - \epsilon B^t$. If r is the real representation of $\mathbb{U}^\pi(n)$ then we have

$$r(\beta) = \begin{pmatrix} A & B \\ B & A \end{pmatrix}, \quad A^t A - B^t B = I_n, \quad A^t B - B^t A = 0$$

for $\beta \in \mathbb{U}^\pi(n)$. Here, I_n denotes the identity matrix of type $(n \times n)$. We consider the group $\mathbb{U}^\pi(n) \times \{1\}$ which consists of matrices α of type $((n+1) \times (n+1))$ such that

$$\alpha = \begin{pmatrix} & & 0 \\ \beta & & \vdots \\ & & 0 \\ 0 \dots 0 & 1 \end{pmatrix} \quad \text{with} \quad \beta \in \mathbb{U}^\pi(n).$$

Then r is extended to a representation of $\mathbb{U}^\pi(n) \times \{1\}$ as

$$r(\alpha) = \begin{pmatrix} A & B & 0 \\ B & A & \vdots \\ & & 0 \\ 0 & \dots 0 & 1 \end{pmatrix}.$$

For $\alpha \in \mathbb{U}^\pi(n) \times \{1\}$, we have $\alpha(\xi) = \xi$, $\alpha \circ \varphi = \varphi \circ \alpha$ and α is an isometry with respect to g, i.e. the matrices of $\mathbb{U}^\pi(n) \times \{1\}$ preserve the structure (ξ, φ, g, η). Hence, $\mathbb{U}^\pi(n) \times \{1\}$ is the structure group of the almost paracontact metric manifolds.

We define a $2n$-dimensional distribution $\mathbb{D}$ by $\mathbb{D}_p = \text{Ker}(\eta_p) = \{x \in T_p M \mid \eta(x) = 0\}$ at each point $p \in M$, and call it a *paracontact distribution* associated with a paracontact structure (φ, ξ, η). According to the decomposition $T_p M = \mathbb{D}_p \oplus span_\mathbb{R}\{\xi(p)\}$ of the tangent space $T_p M$ into orthogonal direct sum, we decompose each vector $x \in T_p M$ as

$$x = hx + vx. \tag{3}$$

Here, its horizontal vector is given by $hx = \varphi^2 x \in \mathbb{D}_p$ and its vertical vector is by $vx = \eta(x) \cdot \xi(p) \in span_\mathbb{R}\{\xi(p)\}$. By the properties (1) the correspondences $h : T_p M \to \mathbb{D}_p$ and $v : T_p M \to span_\mathbb{R}\{\xi(p)\}$ satisfy

$$h(\xi) = 0, \quad h^2 = h, \quad h \circ \varphi = \varphi \circ h = \varphi, \quad v \circ h = h \circ v = 0.$$

The tensor field φ induces an almost paracomplex structure on $\mathbb{D}_p$ at each point $p \in M$, and $(\mathbb{D}_p, \varphi_p, g_{|\mathbb{D}_p})$ admits a $2n$-dimensional almost paracomplex structure (see [7]). Since g is a non-degenerate metric on M, the paracontact distribution $\mathbb{D}$ does not degenerate.

Let ϕ be the fundamental 2-form on $(M, \varphi, \xi, \eta, g)$ defined by

$$\phi(X, Y) = g(\varphi X, Y),$$

and F be the covariant derivative of ϕ with respect to the Levi-Civita connection ∇ of g. That is, the tensor field F of type $(0, 3)$ is defined by

$$F(X, Y, Z) = (\nabla \phi)(X, Y, Z) = (\nabla_X \phi)(Y, Z) = g((\nabla_X \varphi)Y, Z), \quad (4)$$

for $X, Y, Z \in TM$. By use of (1) and (2) we find the tensor field $F = F_M$ has the following properties:

$$\begin{aligned} F(X, Y, Z) &= -F(X, Z, Y), \\ F(X, \varphi Y, \varphi Z) &= F(X, Y, Z) + \eta(Y)F(X, Z, \xi) - \eta(Z)F(X, Y, \xi). \end{aligned} \quad (5)$$

If we put $\mathcal{A}_Y X = (\nabla_X \varphi)Y$ for $X, Y \in TM$, we then have $F(X, Y, Z) = g(\mathcal{A}_Y X, Z)$ for $X, Y, Z \in TM$. Equalities (5) hence turn to

$$\begin{aligned} g(\mathcal{A}_Y X, Z) &= -g(\mathcal{A}_Z X, Y), \\ g(\mathcal{A}_{\varphi Y} X, \varphi Z) &= g(\mathcal{A}_Y X, Z) + \eta(Y)\eta(\mathcal{A}_Z X) - \eta(Z)\eta(\mathcal{A}_Y X). \end{aligned} \quad (6)$$

Next we put $\mathcal{A}'_\xi X = \nabla_X \xi$ for $X \in TM$. Since $g(\xi, \xi) = 1$ we have $g(\nabla_X \xi, \xi) = 0$, hence find $\nabla_X \xi \in \mathbb{D}$. These tensors $\mathcal{A}_\xi$ and $\mathcal{A}'_\xi$ are related by

$$\mathcal{A}_\xi X = -\varphi(\mathcal{A}'_\xi X) \quad (7)$$

because

$$\mathcal{A}_\xi X = (\nabla_X \varphi)\xi = \nabla_X(\varphi \xi) - \varphi(\nabla_X \xi) = -\varphi(\nabla_X \xi) = -\varphi(\mathcal{A}'_\xi X).$$

As immediate consequences of (6) and (7) we see that

$$\begin{aligned} \eta(\mathcal{A}_Y X) &= -g(\mathcal{A}_\xi X, Y) = -g(\mathcal{A}'_\xi X, \varphi Y), \\ \eta(\mathcal{A}_\xi X) &= 0. \end{aligned} \quad (8)$$

We take an arbitrary basis $\{e_1, \ldots, e_{2n}\}$ of $\mathbb{D}_p$, hence $\{e_1, \ldots, e_{2n}, \xi\}$ is a basis of $T_p M$. By use of this basis we define 1-forms θ, θ^*, ω by

$$\begin{aligned} \theta(X) &= g^{ij} F(e_i, e_j, X), \\ \theta^*(X) &= g^{ij} F(e_i, \varphi e_j, X), \\ \omega(X) &= F(\xi, \xi, X), \end{aligned} \quad (9)$$

for $X \in TM$, where the matrix (g^{ij}) of type $(2n, 2n)$ is the inverse matrix of $(g_{ij})_{|\mathbb{D}_p}$. By make use of the tensor $\mathcal{A}_X$ we have

$$\begin{aligned}\theta(X) &= -\text{tr}(\mathcal{A}_X),\\ \theta^*(X) &= \text{tr}(\varphi \circ \mathcal{A}_X) = \text{tr}(\mathcal{A}_X \circ \varphi).\end{aligned} \tag{10}$$

The following two lemmas are well known algebraic results which we apply in the next sections.

Lemma 1.1. *Let $p_1, p_2, \ldots, p_n : V \to V$ be projections of a vector space V satisfying the following conditions:*

$$\sum_{i=1}^{n} p_i = id, \qquad p_i \circ p_j = 0, \ i \neq j.$$

Then we have $V = V_1 \oplus V_2 \oplus \ldots \oplus V_n$, where $V_i = \text{Im}(p_i)$.

Lemma 1.2. *Let $p : V \to V$ be a self-adjoint projection of a vector space V. Then we have $V \cong \text{Ker}(p) \oplus \text{Im}(p)$ and $\text{Ker}(p) \perp \text{Im}(p)$.*

2. Basic classes of almost paracontact metric manifolds

Let V be a $(2n + 1)$-dimensional vector space admitting a metric $g(\cdot, \cdot)$ of signature $(n + 1, n)$, and $\xi \in V$ be a vector satisfying $g(\xi, \xi) = 1$. We define a linear function $\eta : V \to \mathbb{R}$ by $\eta(x) = g(x, \xi)$. We suppose V admits a linear map $\varphi : V \to V$ satisfying $\varphi(\xi) = 0$, $\varphi^2(x) = x - \eta(x)\xi$ and $g(x, \varphi y) = -g(\varphi x, y)$ for all $x, y \in V$. We consider a vector space $\mathcal{F}$ which consists of tensors F of type $(0, 3)$ satisfying

$$\begin{aligned}F(x, y, z) &= -F(x, z, y),\\ F(x, \varphi x, \varphi z) &= F(x, y, z) + \eta(y)F(x, z, \xi) - \eta(z)F(x, y, \xi).\end{aligned} \tag{11}$$

As we explained in section 1, if an almost paracontact metric manifold $(M, \varphi, \xi, \eta, g)$ is given, then at each point $p \in M$ the tangent space $T_p M$ admits a quartet $(\varphi|_{T_p M}, \xi_p, \eta|_{T_p M}, g_p)$ we mentioned above. Since we have a natural tensor field F_M of type $(0, 3)$ on M, we study almost paracontact metric manifolds by properties of the natural tensor fields at each point of base manifolds.

Given $F \in \mathcal{F}$ we define three linear maps θ_F, θ_F^*, $\omega_F : V \to \mathbb{R}$ by

$$\theta_F(x) = g^{ij} F(e_i, e_j, x), \quad \theta_F^*(x) = g^{ij} F(e_i, \varphi e_j, x), \quad \omega_F(x) = F(\xi, \xi, x)$$

with a basis $\{e_1, \ldots, e_{2n}\}$ of $V^0 = \{x \in V \mid \eta(x) = 0\}$. The following is the main theorem in this paper. The outline of its proof is given in the following three sections.

Theorem 2.1. *The vector space $\mathcal{F}$ is decomposed to an orthogonal direct sum $\mathcal{F} = \mathcal{F}_1 \oplus \ldots \oplus \mathcal{F}_{11}$ of eleven subspaces $\mathcal{F}_i \, (i = 1, \ldots, 11)$. These subspaces are invariant under the action of the group $\mathbb{U}^\pi(n) \times \{1\}$ and they are given as follows:*

$$\mathcal{F}_1 = \left\{ F \in \mathcal{F} \,\middle|\, \begin{aligned} &F(x,y,z) \\ &= \frac{1}{2(n-1)}\big\{g(x,\varphi y)\theta_F(\varphi z) - g(x,\varphi z)\theta_F(\varphi y) \\ &\qquad - g(\varphi x, \varphi y)\theta(hz) + g(\varphi x, \varphi z)\theta_F(hy)\big\} \end{aligned} \right\},$$

$$\mathcal{F}_2 = \left\{ F \in \mathcal{F} \,\middle|\, F(\varphi x, \varphi y, z) = -F(x,y,z),\ \theta_F = 0 \right\},$$

$$\mathcal{F}_3 = \left\{ F \in \mathcal{F} \,\middle|\, F(\varphi x, \varphi y, z) = F(x,y,z) \right\},$$

$$\mathcal{F}_4 = \left\{ F \in \mathcal{F} \,\middle|\, F(x,y,z) = \frac{\theta_F(\xi)}{2n}\big\{\eta(y)g(\varphi x, \varphi z) - \eta(z)g(\varphi x, \varphi y)\big\} \right\},$$

$$\mathcal{F}_5 = \left\{ F \in \mathcal{F} \,\middle|\, F(x,y,z) = -\frac{\theta_F^*(\xi)}{2n}\big\{\eta(y)g(x,\varphi z) - \eta(z)g(x,\varphi y)\big\} \right\},$$

$$\mathcal{F}_6 = \left\{ F \in \mathcal{F} \,\middle|\, \begin{aligned} &F(x,y,z) = -F(\varphi x, \varphi y, z) - F(\varphi x, y, \varphi z) \\ &\qquad = -F(y,z,x) + F(z,x,y) - 2F(\varphi x, \varphi y, z), \\ &\theta_F^*(\xi) = 0 \end{aligned} \right\},$$

$$\mathcal{F}_7 = \left\{ F \in \mathcal{F} \,\middle|\, \begin{aligned} &F(x,y,z) = -F(\varphi x, \varphi y, z) - F(\varphi x, y, \varphi z) \\ &\qquad = -F(y,z,x) - F(z,x,y), \\ &\theta_F(\xi) = 0 \end{aligned} \right\},$$

$$\mathcal{F}_8 = \left\{ F \in \mathcal{F} \,\middle|\, \begin{aligned} &F(x,y,z) = F(\varphi x, \varphi y, z) + F(\varphi x, y, \varphi z) \\ &\qquad = -F(y,z,x) + F(z,x,y) + 2F(\varphi x, \varphi y, z) \end{aligned} \right\},$$

$$\mathcal{F}_9 = \left\{ F \in \mathcal{F} \,\middle|\, \begin{aligned} &F(x,y,z) = F(\varphi x, \varphi y, z) + F(\varphi x, y, \varphi z) \\ &\qquad = -F(y,z,x) - F(z,x,y) \end{aligned} \right\},$$

$$\mathcal{F}_{10} = \left\{ F \in \mathcal{F} \,\middle|\, F(x,y,z) = \eta(x)F(\xi, \varphi y, \varphi z) \right\},$$

$$\mathcal{F}_{11} = \left\{ F \in \mathcal{F} \,\middle|\, F(x,y,z) = \eta(x)\{\eta(y)\omega_F(z) - \eta(z)\omega_F(y)\} \right\}.$$

3. Partial decomposition of the space $\mathcal{F}$

The metric $g(\cdot, \cdot)$ on V induces an inner product $\langle \, , \, \rangle$ on $\mathcal{F}$ which is defined by

$$\langle F_1, F_2 \rangle = g^{iq} g^{jr} g^{ks} F_1(f_i, f_j, f_k) F_2(f_q, f_r, f_s) \tag{12}$$

for $F_1, F_2 \in \mathcal{F}$ with a basis $\{f_1, \ldots, f_{2n+1}\}$ of V. Here, (g^{ij}) is the inverse matrix of (g_{ij}) determined by $g_{ij} = g(f_i, f_j)$.

The standard representation of the structure group $\mathbb{U}^\pi(n) \times \{1\}$ in V induces a natural representation λ of $\mathbb{U}^\pi(n) \times \{1\}$ in $\mathcal{F}$ in the following manner:

$$\left(\lambda(\alpha) F \right)(x, y, z) = F\left(\alpha^{-1}x, \alpha^{-1}y, \alpha^{-1}z \right), \tag{13}$$

for $\alpha \in \mathbb{U}^\pi(n) \times \{1\}$ and $F \in \mathcal{F}$. We find $\lambda(\alpha)$ preserves the inner product for each $\alpha \in \mathbb{U}^\pi(n) \times \{1\}$;

$$\langle \lambda(\alpha) F_1, \lambda(\alpha) F_2 \rangle = \langle F_1, F_2 \rangle, \quad F_1, F_2 \in \mathcal{F}. \tag{14}$$

Since the vector space V is decomposed as $V^0 \oplus \mathrm{span}_{\mathbb{R}}\{\xi\}$, we decompose each vector $x \in V$ into $hx + vx$ with $hx \in V^0$ and $vx \in \mathrm{span}_{\mathbb{R}}\{\xi\}$. Using this decomposition we define linear maps $p_i : \mathcal{F} \to \mathcal{F}$ ($i = 1, 2, 3, 4$) analogously as in [4] by

$$\begin{aligned}
p_1(F)(x, y, z) &= F(hx, hy, hz), \\
p_2(F)(x, y, z) &= -\eta(y)F(hx, hz, \xi) + \eta(z)F(hx, hy, \xi), \\
p_3(F)(x, y, z) &= \eta(x)F(\xi, hy, hz), \\
p_4(F)(x, y, z) &= \eta(x)\eta(y)F(\xi, \xi, hz) - \eta(x)\eta(z)F(\xi, \xi, hy).
\end{aligned} \tag{15}$$

One can easily check the following

Lemma 3.1. *The linear maps p_i ($i = 1, 2, 3, 4$) are projections and they have the following properties:*

(i) $\displaystyle\sum_{i=1}^{4} p_i = id_V$;

(ii) $p_i \circ p_j = 0, \quad i \neq j, \quad i, j = 1, 2, 3, 4$;

(iii) *Each p_i ($i = 1, 2, 3, 4$) commutes with the action of $\mathbb{U}^\pi(n) \times \{1\}$ through its representation λ.*

We put $W_i = \mathrm{Im}(p_i) = \{F \in \mathcal{F} \mid F = p_i(F)\}$ for $i = 1, 2, 3, 4$. By the definition of p_i in (15), one can easily check that $\langle F_i, F_j \rangle = 0$ for $F_i \in W_i$, $F_j \in W_j$ ($1 \leq i < j \leq 4$). Hence those subspaces W_1, W_2, W_3 and W_4 are mutually orthogonal. By using Lemma 3.1 and Lemma 1.1 we obtain the following

Proposition 3.1 (Partial decomposition). *The decomposition*

$$\mathcal{F} = W_1 \oplus W_2 \oplus W_3 \oplus W_4$$

is orthogonal and is invariant under the action of the group $\mathbb{U}^\pi(n) \times \{1\}$.

4. The subspace W_1 of $\mathcal{F}$

In this section we decompose the space W_1 into 3 subspaces. Given $F \in \mathcal{F}$ and a vector $y \in V$, we define a linear map $\mathcal{A}_y^F : V \to V$ by $g(\mathcal{A}_y^F x, z) = F(x, y, z)$ for all $x, z \in V$. We also define a linear map $\mathcal{A}_\xi'^F : V \to V$ by $\mathcal{A}_\xi'^F x = -\varphi(\mathcal{A}_\xi^F x)$. By the definitions of p_1 and W_1, we can rewrite the characteristic conditions of W_1 into the following two ways:

$$W_1 = \left\{ F \in \mathcal{F} \mid F(x,y,z) = F(hx, hy, hz) = g(\mathcal{A}_{hy}^F hx, hz) \right\},$$

$$= \left\{ F \in \mathcal{F} \, \middle| \, \begin{array}{l} F(x,y,z) = g(\mathcal{A}_y^F x, z), \\ F(\xi, y, z) = g(\mathcal{A}_y^F \xi, z) = 0, \\ F(x, \xi, z) = g(\mathcal{A}_\xi'^F x, \varphi z) = 0 \end{array} \right\}. \tag{16}$$

Equalities (11) show that

$$\begin{aligned} g(\mathcal{A}_y^F x, z) &= -g(\mathcal{A}_z^F x, y), \\ g(\mathcal{A}_{\varphi y}^F x, \varphi z) &= g(\mathcal{A}_y^F x, z) + \eta(y)\eta(\mathcal{A}_z^F x) - \eta(z)\eta(\mathcal{A}_y^F x). \end{aligned}$$

We also have

$$\theta_F(x) = -\mathrm{tr}(\mathcal{A}_x^F), \quad \theta_F^*(x) = \mathrm{tr}(\varphi \circ \mathcal{A}_x^F) = \mathrm{tr}(\mathcal{A}_x^F \circ \varphi).$$

Hence, when $F \in W_1$, these 1-forms satisfy

$$\theta_F(\xi) = 0, \quad \theta_F^*(z) = \theta_F(\varphi z). \tag{17}$$

We now define linear maps $m_i : W_1 \to W_1$ $(i = 1, 2)$ by

$$\begin{aligned} m_1(F)(x,y,z) &= \frac{1}{2}\{g(\mathcal{A}_y^F x, z) - g(\mathcal{A}_y^F \varphi x, \varphi z)\}, \\ m_2(F)(x,y,z) &= \frac{1}{2}\{g(\mathcal{A}_y^F x, z) + g(\mathcal{A}_y^F \varphi x, \varphi z)\}. \end{aligned} \tag{18}$$

Lemma 4.1. *The linear maps* m_i $(i = 1, 2)$ *are projections and they have the following properties:*

(i) $\displaystyle\sum_{i=1}^2 m_i = id_{W_1}$;

(ii) $m_1 \circ m_2 = m_2 \circ m_1 = 0$;

(iii) *Each* m_i $(i = 1, 2)$ *commutes with the action of* $\mathbb{U}^\pi(n) \times \{1\}$.

Proof. Since W_1 is of the form (16), the definitions of m_1, m_2 in (18) imply

$$
\begin{aligned}
\mathcal{A}_y^{m_1(F)} x &= \frac{1}{2} \left\{ \mathcal{A}_y^F x + \varphi(\mathcal{A}_y^F \varphi x) \right\}, \\
\mathcal{A}_y^{m_2(F)} x &= \frac{1}{2} \left\{ \mathcal{A}_y^F x - \varphi(\mathcal{A}_y^F \varphi x) \right\}.
\end{aligned}
\tag{19}
$$

By (18) we have

$$
m_1(m_1(F))(x, y, z) = \frac{1}{2} \left\{ g\left(\mathcal{A}_y^{m_1(F)} x, z \right) - g\left(\mathcal{A}_y^{m_1(F)} \varphi x, \varphi z \right) \right\},
$$

which shows $m_1(m_1(F))(x, y, z) = m_1(F)(x, y, z)$. Analogously we can also verify that m_2 is a projection.

Next we show their properties.

(i) By (18) we get

$$
\begin{aligned}
(m_1 + m_2)(F)(x, y, z) &= m_1(F)(x, y, z) + m_2(F)(x, y, z) \\
&= g(\mathcal{A}_y^F x, z) = F(x, y, z),
\end{aligned}
$$

hence obtain $m_1 + m_2 = id$.

(ii) Making use of (19) we obtain by direct computations that

$$
\begin{aligned}
m_1(m_2(F))(x, y, z) &= \frac{1}{2} \left\{ g\left(\mathcal{A}_y^{m_2(F)} x, z \right) - g\left(\mathcal{A}_y^{m_2(F)} \varphi x, \varphi z \right) \right\} = 0, \\
m_2(m_1(F))(x, y, z) &= \frac{1}{2} \left\{ g\left(\mathcal{A}_y^{m_1(F)} x, z \right) - g\left(\mathcal{A}_y^{m_1(F)} \varphi x, \varphi z \right) \right\} = 0,
\end{aligned}
$$

which show $m_1 \circ m_2 = m_2 \circ m_1 = 0$.

(iii) We take $\alpha \in \mathbb{U}^\pi(n) \times \{1\}$ and $F \in W_1$. By (13) and (18) we have

$$
m_1\left(\lambda(\alpha) F(x, y, z) \right) = m_1 \left(F\left(\alpha^{-1} x, \alpha^{-1} y, \alpha^{-1} z \right) \right)
$$

$$
= \frac{1}{2} \left\{ g\left(\mathcal{A}_{\alpha^{-1} y}^F \alpha^{-1} x, \alpha^{-1} z \right) - g\left(\mathcal{A}_{\alpha^{-1} y}^F \varphi\left(\alpha^{-1} x \right), \varphi\left(\alpha^{-1} z \right) \right) \right\},
$$

$$
\lambda(\alpha)(m_1(F)(x, y, z)) = \lambda(\alpha) \left(\frac{1}{2} \left\{ g(\mathcal{A}_y^F x, z) - g(\mathcal{A}_y^F \varphi x, \varphi z) \right\} \right)
$$

$$
= \frac{1}{2} \left\{ g\left(\mathcal{A}_{\alpha^{-1} y}^F \alpha^{-1} x, \alpha^{-1} z \right) - g\left(\mathcal{A}_{\alpha^{-1} y}^F \alpha^{-1} \left(\varphi x \right), \alpha^{-1} \left(\varphi z \right) \right) \right\}.
$$

Since the matrix α^{-1} belongs to the structure group and it commutes with φ, we get $m_1 \circ \lambda(\alpha) = \lambda(\alpha) \circ m_1$. Analogously we can verify $m_2 \circ \lambda(\alpha) = \lambda(\alpha) \circ m_2$. $\square$

We put $W_{11} = \mathrm{Im}(m_1)$ and $\mathcal{F}_3 = \mathrm{Im}(m_2)$.

Proposition 4.1. *The decomposition $W_1 = W_{11} \oplus \mathcal{F}_3$ is orthogonal and is invariant under the action of the group $\mathbb{U}^\pi(n) \times \{1\}$.*

Proof. Lemma 4.1 guarantees that $W_1 = W_{11} \oplus \mathcal{F}_3$. By (11), (16) and the definitions of m_1, m_2 in (18), we are able to verify that W_{11} and $\mathcal{F}_3$ are characterized in the following manner:

$$
\begin{aligned}
W_{11} &= \{F \in W_1 \mid \mathcal{A}_y^F \circ \varphi = \varphi \circ \mathcal{A}_y^F\} \\
&= \{F \in W_1 \mid F(x, y, z) = -F(\varphi x, \varphi y, z)\},
\end{aligned}
\tag{20}
$$

$$
\begin{aligned}
\mathcal{F}_3 &= \{F \in W_1 \mid \mathcal{A}_y^F \circ \varphi = -\varphi \circ \mathcal{A}_y^F\} \\
&= \{F \in W_1 \mid F(x, y, z) = F(\varphi x, \varphi y, z)\}.
\end{aligned}
\tag{21}
$$

We take $F_1 \in W_{11}$, $F_2 \in \mathcal{F}_3$. By the above (20) and (21), we have

$$
\begin{aligned}
\langle F_1, F_2 \rangle &= -g^{iq} g^{jr} g^{ks} F_1(\varphi f_i, \varphi f_j, f_k) F_2(\varphi f_q, \varphi f_r, f_s) \\
&= -g^{iq} \varphi_i^\alpha \varphi_q^\lambda g^{jr} \varphi_j^\beta \varphi_r^\mu g^{ks} F_1(f_\alpha, f_\beta, f_k) F_2(f_\lambda, f_\mu, f_s) \\
&= -\frac{4n^2}{(2n+1)^2} g^{\alpha\lambda} g^{\beta\mu} g^{ks} F_1(f_\alpha, f_\beta, f_k) F_2(f_\lambda, f_\mu, f_s) \\
&= -\frac{4n^2}{(2n+1)^2} \langle F_1, F_2 \rangle.
\end{aligned}
$$

Thus we find $\langle F_1, F_2 \rangle = 0$ and obtain that the subspaces W_{11} and $\mathcal{F}_3$ are orthogonal.

Finally, we show that these subspaces are invariant under the action of the group $\mathbb{U}^\pi(n) \times \{1\}$. By Proposition 3.1, we have $\lambda(\alpha)F \in W_1$ for $F \in W_1$ and $\alpha \in \mathbb{U}^\pi(n) \times \{1\}$. Take $F \in W_{11} = \{F \in W_1 \mid F = m_1(F)\}$. We see Lemma 4.1(iii) implies $\lambda(\alpha)F = \lambda(\alpha)(m_1(F)) = m_1(\lambda(\alpha)F)$, hence we find $\lambda(\alpha)F \in W_{11}$. Therefore we obtain W_{11} is invariant under the action of $\mathbb{U}^\pi(n) \times \{1\}$. Since W_1 is also invariant under the action of $\mathbb{U}^\pi(n) \times \{1\}$, we find so is $\mathcal{F}_3$. $\qquad\square$

We define a linear map $m_3 : W_{11} \to W_{11}$ by

$$
\begin{aligned}
m_3(F)&(x, y, z) \\
&= F(x, y, z) - \frac{1}{2(n-1)} \{g(x, \varphi y)\theta_F(\varphi z) - g(x, \varphi z)\theta_F(\varphi y) \\
&\qquad\qquad - g(\varphi x, \varphi y)\theta_F(hz) + g(\varphi x, \varphi z)\theta_F(hy)\}.
\end{aligned}
$$

Lemma 4.2. *The linear map m_3 is a projection and it has the following properties:*

(i) $\langle m_3(F_1), F_2 \rangle = \langle F_1, m_3(F_2) \rangle$, $\quad F_1, F_2 \in W_{11}$;

(ii) *The map m_3 commutes with the action of $\mathbb{U}^\pi(n) \times \{1\}$.*

If we put $\mathcal{F}_1 = \mathrm{Ker}(m_3)$ and $\mathcal{F}_2 = \mathrm{Im}(m_3)$, we find Lemma 4.2 implies the following:

Proposition 4.2. *The decomposition* $W_{11} = \mathcal{F}_1 \oplus \mathcal{F}_2$ *is orthogonal and is invariant under the action of the group* $\mathbb{U}^\pi(n) \times \{1\}$ *These subspaces* $\mathcal{F}_1$, $\mathcal{F}_2$ *are of the form*

$$
\mathcal{F}_1 = \left\{ F \in W_{11} \;\middle|\; \begin{aligned} &F(x,y,z) \\ &= \frac{1}{2(n-1)}\{g(x,\varphi y)\theta_F(\varphi z) - g(x,\varphi z)\theta_F(\varphi y) \\ &\quad - g(\varphi x,\varphi y)\theta_F(hz) + g(\varphi x,\varphi z)\theta_F(hy)\} \end{aligned} \right\},
$$

$$
\mathcal{F}_2 = \{F \in W_{11} \mid F(x,y,z) = -F(\varphi x, \varphi y, z),\ \theta_F = 0\}
$$
$$
= \{F \in W_{11} \mid A_y^F \circ \varphi = \varphi \circ A_y^F,\ \mathrm{tr}(A_y^F \circ \varphi) = 0\}.
$$

Summarizing Proposition 4.1 and Proposition 4.2 up we obtain

Proposition 4.3. *The decomposition* $W_1 = \mathcal{F}_1 \oplus \mathcal{F}_2 \oplus \mathcal{F}_3$ *is orthogonal and is invariant under the action of the group* $\mathbb{U}^\pi(n) \times \{1\}$.

5. The subspace W_2 of $\mathcal{F}$

In this section we decompose the space W_2 into 6 subspaces. By the definitions of A_y^F, A_ξ^F and p_2 we can rewrite the characteristic conditions of W_2 in the following two ways:

$$
W_2 = \left\{ F \in \mathcal{F} \;\middle|\; \begin{aligned} F(x,y,z) &= -\eta(y)F(hx,hz,\xi) \\ &\quad + \eta(z)F(hx,hy,\xi) \\ &= -\eta(y)g(\varphi(A_\xi'^F hx),hz) \\ &\quad + \eta(z)g(\varphi(A_\xi'^F hx),hy) \end{aligned} \right\} \tag{22}
$$

$$
= \left\{ F \in \mathcal{F} \;\middle|\; \begin{aligned} F(x,y,z) &= -\eta(y)g(\varphi(A_\xi'^F x),z) \\ &\quad + \eta(z)g(\varphi(A_\xi'^F x),y), \\ A_\xi'^F \xi &= 0 \end{aligned} \right\}.
$$

By (22) and by definitions of θ_F, θ_F^*, we find for $F \in W_2$ that

$$
\theta_F(\xi) = \mathrm{tr}(A_\xi'^F \circ \varphi), \qquad \theta_F^*(\xi) = -\mathrm{tr}A_\xi'^F.
$$

We define linear maps $q_i : W_2 \to W_2$ $(i = 1, 2)$ by

$$
q_1(F)(x,y,z) = -\frac{1}{2}\eta(y)\left\{g(\varphi(A_\xi'^F x),z) + g(A_\xi'^F(\varphi x),z)\right\}
$$
$$
+ \frac{1}{2}\eta(z)\left\{g(\varphi(A_\xi'^F x),y) + g(A_\xi'^F(\varphi x),y)\right\},
$$

$$q_2(F)(x,y,z) = -\frac{1}{2}\eta(y)\left\{g(\varphi(\mathcal{A}'^F_\xi x), z) - g(\mathcal{A}'^F_\xi(\varphi x), z)\right\}$$
$$+ \frac{1}{2}\eta(z)\left\{g(\varphi(\mathcal{A}'^F_\xi x), y) - g(\mathcal{A}'^F_\xi(\varphi x), y)\right\}.$$

Lemma 5.1. *The linear maps* q_i $(i = 1, 2)$ *are projections and they have the following properties:*

(i) $\sum_{i=1}^{2} q_i = id_{W_2}$;

(ii) $q_1 \circ q_2 = q_2 \circ q_1 = 0$;

(iii) *Each* q_i $(i = 1, 2)$ *commutes with the action of* $\mathbb{U}^\pi(n) \times \{1\}$.

We put $W' = \text{Im}(q_1)$ and $W'' = \text{Im}(q_2)$. These subspaces are represented as

$$W' = \left\{F \in W_2 \ \Big| \ \mathcal{A}'^F_\xi \circ \varphi = \varphi \circ \mathcal{A}'^F_\xi\right\}$$
$$= \{F \in W_2 \ | \ F(x,y,z) = -F(\varphi x, \varphi y, z) - F(\varphi x, y, \varphi z)\}, \tag{23}$$

$$W'' = \left\{F \in W_2 \ \Big| \ \mathcal{A}'_\xi \circ \varphi = -\varphi \circ \mathcal{A}'^F_\xi\right\}$$
$$= \{F \in W_2 \ | \ F(x,y,z) = F(\varphi x, \varphi y, z) + F(\varphi x, y, \varphi z)\}. \tag{24}$$

We can check that W' and W'' are orthogonal by these characterizations. Taking into account Lemma 5.1 and Lemma 1.1 we obtain the following:

Proposition 5.1. *The decomposition* $W_2 = W' \oplus W''$ *is orthogonal and is invariant under the action of the group* $\mathbb{U}^\pi(n) \times \{1\}$.

We define linear maps $r_i : W' \to W'$ $(i = 1, 2)$ by

$$r_1(F)(x,y,z) = -\frac{1}{2}\eta(y)\left\{g(\varphi(\mathcal{A}'^F_\xi x), z) + g(\varphi x, \mathcal{A}'^F_\xi z)\right\}$$
$$+ \frac{1}{2}\eta(z)\left\{g(\varphi(\mathcal{A}'^F_\xi x), y) + g(\varphi x, \mathcal{A}'^F_\xi y)\right\},$$
$$r_2(F)(x,y,z) = -\frac{1}{2}\eta(y)\left\{g(\varphi(\mathcal{A}'^F_\xi x), z) - g(\varphi x, \mathcal{A}'^F_\xi z)\right\}$$
$$+ \frac{1}{2}\eta(z)\left\{g(\varphi(\mathcal{A}'^F_\xi x), y) - g(\varphi x, \mathcal{A}'^F_\xi y)\right\}.$$

Lemma 5.2. *The linear maps* r_i $(i = 1, 2)$ *are projections and they have the following properties:*

(i) $\sum_{i=1}^{2} r_i = id_{W'}$;

(ii) $r_1 \circ r_2 = r_2 \circ r_1 = 0$;

(iii) *Each r_i $(i = 1, 2)$ commutes with the action of $\mathbb{U}^{\pi}(n) \times \{1\}$.*

We put $W_1' = \mathrm{Im}(r_1)$ and $W_2' = \mathrm{Im}(r_2)$. These subspaces are represented as follows:

$$W_1' = \left\{ F \in W' \mid g(\mathcal{A}_{\xi}'^F \cdot, \cdot) = g(\cdot, \mathcal{A}_{\xi}'^F \cdot) \right\}$$

$$= \left\{ F \in W' \mid \begin{array}{c} F(x, y, z) = -F(y, z, x) + F(z, x, y) \\ - 2F(\varphi x, \varphi y, z) \end{array} \right\}, \quad (25)$$

$$W_2' = \left\{ F \in W' \mid g(\mathcal{A}_{\xi}'^F \cdot, \cdot) = -g(\cdot, \mathcal{A}_{\xi}'^F \cdot) \right\}$$

$$= \left\{ F \in W' \mid F(x, y, z) = -F(y, z, x) - F(z, x, y) \right\}. \quad (26)$$

We can check that W_1' and W_2' are orthogonal by these characterizations. Taking into account Lemma 5.2 and Lemma 1.1 we obtain the following:

Proposition 5.2. *The decomposition $W' = W_1' \oplus W_2'$ is orthogonal and is invariant under the action of the group $\mathbb{U}^{\pi}(n) \times \{1\}$.*

We define a linear map $s : W_1' \to W_1'$ by

$$s(F)(x, y, z) = F(x, y, z) + \frac{\theta_F^*(\xi)}{2n} \left\{ \eta(y)g(x, \varphi z) - \eta(z)g(x, \varphi y) \right\}.$$

Lemma 5.3. *The linear map s is a projection and it has the following properties:*

(i) $\langle s(F_1), F_2 \rangle = \langle F_1, s(F_2) \rangle$, $F_1, F_2 \in W_1'$;

(ii) *s commutes with the action of $\mathbb{U}^{\pi}(n) \times \{1\}$.*

Putting $\mathcal{F}_5 = \mathrm{Ker}(s)$ and $\mathcal{F}_6 = \mathrm{Im}(s)$, we obtain the following with the aid of Lemma 5.3.

Proposition 5.3. *The decomposition $W_1' = \mathcal{F}_5 \oplus \mathcal{F}_6$ is orthogonal and is invariant under the action of the group $\mathbb{U}^{\pi}(n) \times \{1\}$. These subspaces are represented as*

$$\mathcal{F}_5 = \left\{ F \in W_1' \mid F(x, y, z) = -\frac{\theta_F^*(\xi)}{2n} \left\{ \eta(y)g(x, \varphi z) - \eta(z)g(x, \varphi y) \right\} \right\}$$

$$= \left\{ F \in W_1' \;\middle|\; \mathcal{A}_\xi'^F = \frac{\mathrm{tr}\mathcal{A}_\xi'^F}{2n} \, id \right\},$$

$$\mathcal{F}_6 = \{ F \in W_1' \mid \theta_F^*(\xi) = 0 \} = \left\{ F \in W_1' \;\middle|\; \mathrm{tr}\mathcal{A}_\xi'^F = 0 \right\}.$$

We define a linear map $t : W_2' \to W_2'$ by

$$t(F)(x,y,z) = F(x,y,z) - \frac{\theta_F(\xi)}{2n} \{ \eta(y)g(\varphi x, \varphi z) - \eta(z)g(\varphi x, \varphi y) \}.$$

Lemma 5.4. *The linear map t is a projection and it has the following properties*:

(i) $\langle t(F_1), F_2 \rangle = \langle F_1, t(F_2) \rangle, \quad F_1, F_2 \in W_2'$;
(ii) *The map t commutes with the action of* $\mathbb{U}^\pi(n) \times \{1\}$.

If we put $\mathcal{F}_4 = \mathrm{Ker}(t)$ and $\mathcal{F}_7 = \mathrm{Im}(t)$, we obtain the following by Lemma 5.4.

Proposition 5.4. *The decomposition $W_2' = \mathcal{F}_4 \oplus \mathcal{F}_7$ is orthogonal and is invariant under the action of the group $\mathbb{U}^\pi(n) \times \{1\}$. These subspaces are represented as*

$$\mathcal{F}_4 = \left\{ F \in W_2' \;\middle|\; F(x,y,z) = \frac{\theta_F(\xi)}{2n} \{ \eta(y)g(\varphi x, \varphi z) - \eta(z)g(\varphi x, \varphi y) \} \right\}$$

$$= \left\{ F \in W_2' \;\middle|\; \mathcal{A}_\xi'^F \circ \varphi = \frac{\mathrm{tr}(\mathcal{A}_\xi'^F \circ \varphi)}{2n} \, id \right\},$$

$$\mathcal{F}_7 = \{ F \in W_2' \mid \theta_F(\xi) = 0 \} = \left\{ F \in W_2' \;\middle|\; \mathrm{tr}(\mathcal{A}_\xi'^F \circ \varphi) = 0 \right\}.$$

Next we decompose the subspace W'' of W_2. We define linear maps $l_i : W'' \to W''$ $(i = 1, 2)$ by

$$l_1(F)(x,y,z) = -\frac{1}{2}\eta(y) \left\{ g(\varphi(\mathcal{A}_\xi'^F x), z) - g(\varphi x, \mathcal{A}_\xi'^F z) \right\}$$
$$+ \frac{1}{2}\eta(z) \left\{ g(\varphi(\mathcal{A}_\xi'^F x), y) - g(\varphi x, \mathcal{A}_\xi'^F y) \right\},$$

$$l_2(F)(x,y,z) = -\frac{1}{2}\eta(y) \left\{ g(\varphi(\mathcal{A}_\xi'^F x), z) + g(\varphi x, \mathcal{A}_\xi'^F z) \right\}$$
$$+ \frac{1}{2}\eta(z) \left\{ g(\varphi(\mathcal{A}_\xi'^F x), y) + g(\varphi x, \mathcal{A}_\xi'^F y) \right\}.$$

Lemma 5.5. *These linear maps l_i $(i = 1, 2)$ are projections and they have the following properties*:

(i) $\displaystyle\sum_{i=1}^{2} l_i = id_{W''}$;

(ii) $l_1 \circ l_2 = l_2 \circ l_1 = 0$;

(iii) *Each l_i ($i = 1, 2$) commute with the action of $\mathbb{U}^\pi(n) \times \{1\}$.*

We put $\mathcal{F}_9 = \operatorname{Im}(l_1)$ and $\mathcal{F}_8 = \operatorname{Im}(l_2)$. These subspaces are represented as follows:

$$
\mathcal{F}_8 = \left\{ F \in W'' \;\middle|\; g(\mathcal{A}'^F_\xi \,\cdot\,, \,\cdot\,) = -g(\,\cdot\,, \mathcal{A}'^F_\xi \,\cdot\,) \right\}
$$

$$
= \left\{ F \in W'' \;\middle|\; \begin{array}{l} F(x,y,z) = -F(y,z,x) + F(z,x,y) \\ \qquad\qquad + 2F(\varphi x, \varphi y, z) \end{array} \right\}, \tag{27}
$$

$$
\mathcal{F}_9 = \left\{ F \in W'' \;\middle|\; g(\mathcal{A}'^F_\xi \,\cdot\,, \,\cdot\,) = g(\,\cdot\,, \mathcal{A}'^F_\xi \,\cdot\,) \right\} \tag{28}
$$

$$
= \left\{ F \in W'' \;\middle|\; F(x,y,z) = -F(y,z,x) - F(z,x,y) \right\}.
$$

We can check that $\mathcal{F}_8$ and $\mathcal{F}_9$ are orthogonal by these characterizations. Taking into account Lemma 5.5 and Lemma 1.1 we obtain the following:

Proposition 5.5. *The decomposition $W'' = \mathcal{F}_8 \oplus \mathcal{F}_9$ is orthogonal and is invariant under the action of the group $\mathbb{U}^\pi(n) \times \{1\}$.*

Finally, we put $\mathcal{F}_{10} = W_3$ and $\mathcal{F}_{11} = W_4$. They are represented as

$$
\mathcal{F}_{10} = \left\{ F \in \mathcal{F} \;\middle|\; F(x,y,z) = \eta(x) F(\xi, hy, hz) = \eta(x) g(\mathcal{A}^F_{hy}\xi, hz) \right\} \tag{29}
$$

$$
= \left\{ F \in \mathcal{F} \;\middle|\; F(x,y,z) = \eta(x) F(\xi, \varphi y, \varphi z) = \eta(x) g(\mathcal{A}^F_{\varphi y}\xi, \varphi z) \right\},
$$

$$
\mathcal{F}_{11} = \left\{ F \in \mathcal{F} \;\middle|\; F(x,y,z) = \eta(x)\eta(y)\omega_F(z) - \eta(x)\eta(z)\omega_F(y) \right\} \tag{30}
$$

$$
= \left\{ F \in \mathcal{F} \;\middle|\; \begin{array}{l} F(x,y,z) \\ \quad = \eta(x)\{\eta(y) g(\mathcal{A}'^F_\xi \xi, \varphi z) - \eta(z) g(\mathcal{A}'^F_\xi \xi, \varphi y)\} \end{array} \right\}.
$$

Taking into account Propositions 3.1, 4.3, 5.1, 5.2, 5.3, 5.4, 5.5 and the representations (29), (30), we obtain the assertion of Theorem 2.1.

Corresponding to the decomposition of the space $\mathcal{F}$ in Theorem 2.1, we give 11 classes of almost paracontact metric manifolds. An almost paracontact metric manifold M is said to be in the class $\mathcal{F}_i$ ($i = 1, \ldots, 11$) (or $\mathcal{F}_i$-manifold) if at each $p \in M$ the function $F_p : T_pM \times T_pM \times T_pM \to \mathbb{R}$ belongs to the subspace $\mathcal{F}_i$. We say an almost paracontact metric manifold belongs to the class $\mathcal{F}_0$ if it corresponds to the null tensor (i.e. $F(X,Y,Z) = 0$). We note that each class $\mathcal{F}_i$ $(i = 1, \ldots, 11)$ contains this special class. We should note that we do not classify all almost paracontact manifolds. When an almost paracontact manifold M is not homogeneous, the class of F_p may depend on the choice of point $p \in M$.

6. Examples

Let V be a 5-dimensional real vector space. We consider the structure of the Lie algebra on V defined by the brackets $[E_i, E_j] = C_{ij}^k.E_k$, where $\{E_1, E_2, E_3, E_4, E_5\}$ is a basis of V and $C_{ij}^k \in \mathbb{R}$. Let G be the associated real connected Lie group and $\mathfrak{g}$ be the real Lie algebra of G with a basis of left invariant vector fields $\{X_1, X_2, X_3, X_4, X_5\}$ related to $\{E_1, \ldots, E_5\}$. These vector fields satisfy the Jacobi identity

$$\underset{X_i, X_j, X_k}{\mathfrak{S}} \left[[X_i, X_j], X_k\right] = 0. \tag{31}$$

We define a tensor field φ of type $(1,1)$ by the condition

$$\varphi X_i = X_{2+i}, \quad \varphi X_{2+i} = X_i \ (i = 1, 2), \quad \varphi X_5 = 0, \tag{32}$$

and metric g and 1-form η on G by

$$\begin{aligned}
g(X_i, X_i) &= -g(X_{2+i}, X_{2+i}) = g(X_5, X_5) = 1 \ (i = 1, 2), \\
g(X_j, X_k) &= 0, && (j \neq k, \ j, k = 1, 2, 3, 4, 5), \\
\eta(X_j) &= g(X_j, \xi), && (j = 1, 2, 3, 4, 5).
\end{aligned} \tag{33}$$

We set $\xi = X_5$. Then the quartet (φ, ξ, η, g) turns to be an almost paracontact metric structure on G, hence $(G, \varphi, \xi, \eta, g)$ is a 5-dimensional almost paracontact metric manifold.

Example 6.1. We define the commutators of the basis vector fields of $\mathfrak{g}$ in the following way:

$$\begin{aligned}
[X_1, X_2] &= (\lambda - 2c)X_1, \\
[X_1, \varphi X_1] &= [X_2, \varphi X_2] = 2(c - \lambda)\varphi X_2, \\
[X_2, \varphi X_1] &= 2aX_1 - \lambda \varphi X_1, \\
[X_1, \varphi X_2] &= [X_1, \xi] = [X_2, \xi] = [\varphi X_1, \varphi X_2] \\
&= [\varphi X_1, \xi] = [\varphi X_2, \xi] = 0,
\end{aligned} \tag{34}$$

where a, c, $\lambda \in \mathbb{R}$. Then the manifold $(G, \varphi, \xi, \eta, g)$ is an almost paracontact metric manifold in the class $\mathcal{F}_3$.

This can be checked in the following way. Since the metric g is left-invariant, we have

$$2g(\nabla_X Y, Z) = g([X,Y], Z) + g([Z,X], Y) + g([Z,Y], X) \qquad (35)$$

for the Levi-Civita connection ∇ of g. Thus we get the components of the tensor F. The non-zero components of F are

$$F_{112} = F_{314} = -F_{323} = a, \quad F_{114} = F_{312} = -F_{123} = c,$$

where $F_{ijk} = F(X_i, X_j, X_k)$. By direct computation we obtain that $F(X, Y, Z) = F(\varphi X, \varphi Y, Z)$. Hence we find $F \in \mathcal{F}_3$.

Example 6.2. We define the commutators of the basis vector fields by

$$\begin{aligned}
[X_1, \xi] &= aX_1, \quad [X_2, \xi] = cX_2, \\
[\varphi X_1, \xi] &= -a\varphi X_1, \quad [\varphi X_2, \xi] = -c\varphi X_2, \\
[X_1, X_2] &= [X_1, \varphi X_1] = [X_1, \varphi X_2] \\
&= [\varphi X_1, \varphi X_2] = [X_2, \varphi X_1] = [X_2, \varphi X_2] = 0,
\end{aligned} \qquad (36)$$

where $a,\, c \in \mathbb{R}$. Then the manifold $(G, \varphi, \xi, \eta, g)$ is an almost paracontact metric manifold in the class $\mathcal{F}_9$.

This can be checked in the following way. Since the metric g is left-invariant, we find that the non-zero components of F are

$$F_{135} = F_{315} = -a, \quad F_{245} = F_{425} = -c.$$

By direct computation we obtain that

$$\begin{aligned}
F(X, Y, Z) &= F(\varphi X, \varphi Y, Z) + F(\varphi X, Y, \varphi Z) \\
&= -F(Y, Z, X) - F(Z, X, Y).
\end{aligned}$$

Hence we find $F \in \mathcal{F}_9$.

Remark 6.1. The least possible dimension of an almost paracontact metric manifold is 3. If we regard the conditions of the classes $\mathcal{F}_i$ $(i = 1, \ldots, 11)$ written by the linear operators $\mathcal{A}_Y$ and $\mathcal{A}'_\xi$ we can verify that an almost paracontact metric manifold of dimension 3 cannot belong to the classes $\mathcal{F}_1, \mathcal{F}_2, \mathcal{F}_3$ and $\mathcal{F}_6$.

Acknowledgments

The second author acknowledges the support from the European Operational programm HRD through contract BGO051PO001/07/3.3-02/53 with the Bulgarian Ministry of Education. He was also partially supported by Contract 198/2010 with "St. Kl. Ohridski" University of Sofia.

The first author was partially supported by the Scientific researches fund of "St. Cyril and St. Methodius" University of Veliko Tarnovo under contract RD-491-08 from 27.06.2008.

References

1. C. Bejan, *A classification of the almost parahermitian manifolds*, Proc. Conference on Diff. Geom. and Appl., Dubrovnik 1988, 23–27.

2. I. Sato, *On a structure similar to the almost contact structure*, Tensor. New Series, 30(1976), no. 3, , 219–224.

3. D. E. Blair, *Riemannian Geometry of Contact and Symplectic Manifolds*, Progress in Mathematics 203, Birkhäuser, Boston, Mass, USA, 2002.

4. G. Ganchev, V. Mihova, K. Gribachev, *Almost Contact Manifolds with B-Metric*, Mathematica Balkanica 7(1993), 261–277.

5. A. Gray, L.M. Hevella, *The sixteen classes of almost Hermitian manifolds and their linear invariants*, Ann. di Mat. 123(1980), 35–58.

6. S. Kaneyuki, M. Konzai, *Paracomplex structures and affine symmetric spaces*, Tokyo J. Math. 8(1985), 301–318.

7. S. Kaneyuki, F. L. Willams, *Almost paracontact and parahodge structures on manifolds*, Nagoya Math. J. 99(1985), 173–187.

8. M. Manev, M. Staikova, *On almost paracontact Riemannian manifolds of type* (n, n), J. Geom., 72(2001), 108–14.

9. A.M. Naveira, *A classification of Riemannian almost-product manifolds*, Rend. Mat. Appl. (7) 3(1983), 577–592.

10. S. Zamkovoy, *Canonical connections on para-contact manifolds*, arXiv:0707.1787.

Received January 10, 2011
Revised   April 29, 2011

Proceedings of the 2nd International
Colloquium on Differential Geometry
and its Related Fields
Veliko Tarnovo, September 6–10, 2010

NOTES ON GEOMETRY OF q-NORMAL DISTRIBUTIONS

Daiki TANAYA

Department of Computer Science and Engineering,
Graduate School of Engineering, Nagoya Institute of Technology,
Nagoya, Aichi 466-8555 Japan

Masaru TANAKA

Department of Applied Mathematics,
Faculty of Science, Fukuoka University,
8-19-1 Nakamura, Johnan-ku, Fukuoka 814-0180 Japan
E-mail: sieger@math.sci.fukuoka-u.ac.jp

Hiroshi MATSUZOE

Department of Computer Science and Engineering,
Graduate School of Engineering, Nagoya Institute of Technology,
Nagoya, Aichi 466-8555 Japan
E-mail: matsuzoe@nitech.ac.jp

A q-normal distribution is a natural generalization of the standard normal distribution, which is introduced in Tsallis nonextensive statistical mechanics. In the case $1 < q < 3$, the set of q-normal distributions can be regarded as a two dimensional manifold, and the manifold naturally has two kinds of Riemannian metrics, the Fisher metric and the q-Fisher metric. In this paper, geometry of these Riemannian metrics is discussed, and it is shown that these two metrics are conformally equivalent. In addition, it is shown that the set of q-normal distributions with the Fisher metric is a space of constant negative sectional curvature $\kappa = -q/(3-q)$.

Keywords: q-normal distribution, Information geometry, Tsallis statistics, Statistical manifold, Conformal metric.

Introduction

A normal distribution is one of elementary probability distributions, which is widely used in mathematical sciences. It is known that a normal distribution can be obtained from the maximization of Boltzmann-Gibbs-Shannon entropy, and that the set of normal distributions can be regarded as a two dimensional Riemannian manifold. The Riemannian metric naturally defined on this manifold is called the Fisher metric.

For a fixed number $-\infty < q < 3$, a q-normal distribution is a natural generalization of a normal distribution, which is introduced in nonextensive statistical mechanics [4, 12]. A q-normal distribution can be obtained from the maximization of a generalized nonextensive entropy, called the Tsallis entropy. In the case $1 < q < 3$, the set of q-normal distributions can be regarded as a two dimensional manifold, and this manifold naturally has two kinds of Riemannian metrics. The one is the standard Fisher metric, and the other is a Hessian metric, called the q-Fisher metric (cf. [10]).

In this paper, we consider geometry of the set of q-normal distributions. In particular, we discuss relations between the Fisher metric and the q-Fisher metric. We then show that these two metrics are conformally equivalent.

The set of normal distributions with the Fisher metric is a space of constant negative sectional curvature $\kappa = -1/2$ [1, 3]. We show that the set of q-normal distributions is a space of constant negative sectional curvature $\kappa = -q/(3 - q)$.

1. Geometry of normal distributions

In this section, we review geometry of normal distributions. For more details, see [1, 3].

Let S be the set of normal distributions, that is,

$$S = \left\{ p(x; \mu, \sigma) \,\middle|\, p(x; \mu, \sigma) = \frac{1}{\sqrt{2\pi}\sigma} \exp\left[-\frac{(x - \mu)^2}{2\sigma^2} \right] \right\},$$

where x is a random variable, μ and σ are parameters. We call S the *normal family*. For simplicity, set $(\xi^1, \xi^2) = (\mu, \sigma)$. In this parametrization, the parameter space Ξ is the upper half plane:

$$\Xi = \{(\xi^1, \xi^2)\} = \{(\mu, \sigma)| -\infty < \mu < \infty, 0 < \sigma < \infty\}.$$

Since a normal distribution $p(x; \mu, \sigma)$ and this (μ, σ)-parameterization have one-to-one correspondence, S can be regarded as a manifold with a local coordinate system $\{\mu, \sigma\}$ [1].

For a function $f(x)$, denote by $E_\xi[f(x)]$ the expectation of $f(x)$ with respect to $p(x; \xi) = p(x; \mu, \sigma)$. Set $l_\xi = \ln p(x; \xi)$. For the normal family S, we define a Riemannian metric, called the *Fisher metric* $g^F = (g^F_{ij})$, by the following formula:

$$g^F_{ij}(\xi) := \int_\Omega \left(\frac{\partial}{\partial \xi^i} \ln p(x; \xi) \right) \left(\frac{\partial}{\partial \xi^j} \ln p(x; \xi) \right) p(x; \xi) dx \qquad (1)$$
$$= E_\xi \left[\frac{\partial l_\xi}{\partial \xi^i} \frac{\partial l_\xi}{\partial \xi^j} \right].$$

In the (μ, σ)-parametrization, the Fisher metric is

$$g^F(\xi) = \frac{1}{\sigma^2} \begin{pmatrix} 1 & 0 \\ 0 & 2 \end{pmatrix}.$$

Hence we have the sectional curvature $\kappa = -1/2$, and the manifold S with the Fisher metric is a space of constant negative curvature [1, 3].

We change parameters as

$$\theta^1 = \frac{\mu}{\sigma^2}, \quad \theta^2 = -\frac{1}{2\sigma^2},$$

and set

$$F_1(x) = x, \quad F_2(x) = x^2,$$

$$\psi(\theta) = \frac{\mu^2}{2\sigma^2} + \ln(\sqrt{2\pi}\sigma) = -\frac{(\theta^1)^2}{4\theta^2} + \frac{1}{2}\ln\left(-\frac{\pi}{\theta^2}\right).$$

Then we obtain

$$p(x; \mu, \sigma) = \frac{1}{\sqrt{2\pi}\sigma} \exp\left[-\frac{(x-\mu)^2}{2\sigma^2}\right]$$

$$= \exp\left[\frac{\mu}{\sigma^2}x - \frac{1}{2\sigma^2}x^2 - \frac{\mu^2}{2\sigma^2} - \ln(\sqrt{2\pi}\sigma)\right]$$

$$= \exp\left[x\theta^1 + x^2\theta^2 - \psi(\theta)\right].$$

This implies that the normal family is an *exponential family*. The coordinate system $\{\theta^i\}$ is called the *natural parameters*.

We note that the coordinate system $\{\theta^i\}$ is an affine coordinate system of the exponential connection $\nabla^{(e)}$ which is a flat affine connection naturally defined on the normal family [1].

The function $\psi(\theta)$ is a potential of the Fisher metric g^F in the $\{\theta^i\}$-coordinates. In fact,

$$\frac{\partial^2\psi}{\partial\theta^i\partial\theta^j} = \frac{\partial^2\psi}{\partial\theta^i\partial\theta^j}\int_{-\infty}^{\infty} p(x;\theta)dx$$

$$= -\int_{-\infty}^{\infty}\left(\frac{\partial^2}{\partial\theta^i\partial\theta^j}\ln p(x;\theta)\right)p(x;\theta)dx$$

$$= \int_{-\infty}^{\infty}\left(\frac{\partial}{\partial\theta^j}\ln p(x;\theta)\right)\left(\frac{\partial}{\partial\theta^i}p(x;\theta)\right)dx$$

$$= \int_{-\infty}^{\infty}\left(\frac{\partial}{\partial\theta^j}\ln p(x;\theta)\right)\left(\frac{\partial}{\partial\theta^i}\ln p(x;\theta)\right)p(x;\theta)dx$$

$$= g_{ij}^F.$$

Differentiating twice the potential function ψ, we immediately obtain the following proposition.

Proposition 1.1. *For the normal family S, the Fisher metric in natural parameters is given as follows:*

$$g^F(\theta) = -\frac{1}{2\theta^2} \begin{pmatrix} 1 & -\dfrac{\theta^1}{\theta^2} \\ -\dfrac{\theta^1}{\theta^2} & \dfrac{(\theta^1)^2 - \theta^2}{(\theta^2)^2} \end{pmatrix}.$$

See also Shima[7]. Similar arguments are given in §6. The Fisher metric g^F on the normal family S is a *Hessian metric* since g^F is given as a Hessian of the potential function with respect to an affine coordinate system.

We define the *expectation parameters* (η_1, η_2) by

$$\eta_1 = E_\xi[F_1(x)] = \mu,$$
$$\eta_2 = E_\xi[F_2(x)] = \mu^2 + \sigma^2.$$

It is known that $\{\eta_i\}$ is the *dual coordinate system* of $\{\theta^i\}$ with respect to the Fisher metric g^F [1], that is, the following equation holds:

$$g^F\left(\frac{\partial}{\partial\theta^i}, \frac{\partial}{\partial\eta_j}\right) = \delta_i^j.$$

The Fisher metric in the $\{\eta_i\}$-coordinates and its potential function are given by

$$g^F(\eta) = \frac{1}{(\eta_2 - (\eta_1)^2)^2} \begin{pmatrix} \eta_2 + (\eta_1)^2 & -\eta_1 \\ -\eta_1 & \dfrac{1}{2} \end{pmatrix},$$

$$\phi(\eta) = -\frac{1 + \ln 2\pi}{2} - \frac{1}{2}\ln\left(\eta_2 - (\eta_1)^2\right).$$

The parameter changes between the $\{\theta^i\}$- and the $\{\eta^i\}$-coordinates are given as follows:

$$\eta_1 = -\frac{\theta^1}{2\theta^2} = \frac{\partial}{\partial\theta^1}\psi(\theta), \qquad \eta_2 = \frac{(\theta^1)^2}{4(\theta^2)^2} - \frac{1}{2\theta^2} = \frac{\partial}{\partial\theta^2}\psi(\theta),$$

$$\theta^1 = \frac{\eta_1}{\eta_2 - (\eta_1)^2} = \frac{\partial}{\partial\eta_1}\phi(\eta), \qquad \theta^2 = -\frac{1}{2(\eta_2 - (\eta_1)^2)} = \frac{\partial}{\partial\eta_2}\phi(\eta).$$

We note that the η-potential function $\phi(\eta)$ is given by the expectation of the log-likelihood function, that is, the following equation holds:

$$\phi(\eta) = E_\eta[\ln p(x; \eta)].$$

2. q-normal distributions

In this section, we summerize foundations of q-normal distributions. Further details of q-normal distributions, see [8] and [9], for example.

For a fixed real number $q(-\infty < q < 3)$, we define a q-*normal distribution* by the following formula:

$$p_q(x; \mu, \sigma) = \frac{1}{Z_{q,\sigma}} \left[1 - \frac{1-q}{3-q} \frac{(x-\mu)^2}{\sigma^2} \right]_+^{\frac{1}{1-q}},$$

where $[*]_+ = \max\{0, *\}$, x is a random variable, μ and σ are parameters $-\infty < \mu < \infty, 0 < \sigma < \infty$. Here $Z_{q,\sigma}$ is the normalization defined by

$$Z_{q,\sigma} = A_q \sigma,$$

$$A_q = \begin{cases} \sqrt{\frac{3-q}{1-q}} \operatorname{Beta}\left(\frac{2-q}{1-q}, \frac{1}{2}\right), & -\infty < q < 1, \\ \sqrt{\frac{3-q}{q-1}} \operatorname{Beta}\left(\frac{3-q}{2(q-1)}, \frac{1}{2}\right), & 1 \le q < 3, \end{cases}$$

where $\operatorname{Beta}(s, t)$ is the beta-function defined by

$$\operatorname{Beta}(s, t) = \int_0^1 x^{s-1}(1-x)^{t-1} dx, \quad s > 0, \ t > 0.$$

In the case of $q = 1$, we define the normalization A_1 by

$$A_1 = \sqrt{2\pi} = \lim_{q \to 1} A_q = \begin{cases} \lim_{q \to 1-0} \sqrt{\frac{3-q}{1-q}} \operatorname{Beta}\left(\frac{2-q}{1-q}, \frac{1}{2}\right), \\ \lim_{q \to 1+0} \sqrt{\frac{3-q}{q-1}} \operatorname{Beta}\left(\frac{3-q}{2(q-1)}, \frac{1}{2}\right). \end{cases}$$

We remark that the q-normal distribution is the normal distribution if q equals one. If $q > 1$, set $\nu = (3-q)/(q-1)$ and $\lambda = 1/\sigma^2$. Then the q-normal distribution is the three-parameter version of Student's t-distribution

$$p_\nu(x; \mu, \lambda) = \frac{\Gamma(\frac{\nu+1}{2})}{\Gamma(\frac{\nu}{2})} \sqrt{\frac{\lambda}{\pi\nu}} \left[1 + \frac{\lambda(x-\mu)^2}{\nu} \right]^{-\frac{\nu+1}{2}}$$

(see [2]). Here the parameter ν is called the *degree of freedom*, and λ is called the *precision*. In particular, if $q = 2$, then the distribution is the Cauchy distribution.

The q-normal distribution is the semi-circle distribution if $q = -1$, and that the distribution goes to the uniform distribution if $q \to -\infty$.

We also remark that the q-normal distribution does not have the mean of x if $q \ge 2$, and does not have the variance if $q \ge 5/3$. The first and the second moments diverge such q-values.

3. The escort distribution and τ-transformation

For a q-normal distribution, we can define two kinds of expectations. The one is the standard expectation, the other is the q-expectation.

Let $p(x)$ be a probability distribution on R. For a fixed number q, we define the q-escort distribution $P_q(x)$ of $p(x)$ by

$$P_q(x) := \frac{1}{\Omega_q(p)} p(x)^q,$$

where $\Omega_q(p)$ is the normalization defined by

$$\Omega_q(p) := \int_{-\infty}^{\infty} p(x)^q dx.$$

Suppose that $f(x)$ is a random variable on R. The expectation with respect to the q-escort distribution is called the q-expectation, that is,

$$E_{q,p}[f(x)] := \int_{-\infty}^{\infty} f(x)P_q(x)dx = \frac{1}{\Omega_q(p)} \int_{-\infty}^{\infty} f(x)p(x)^q dx.$$

The following proposition was obtained in [11].

Proposition 3.1. Let $p_q(x; \mu, \sigma)$ be a q-normal distribution. Then the q-escort distribution of $p_q(x; \mu, \sigma)$ is given as follows:

$$P_q(x; \mu, \sigma) = \frac{2}{3-q} \frac{1}{Z_{q,\sigma}} \left[1 - \frac{1-q}{3-q} \frac{(x-\mu)^2}{\sigma^2} \right]_+^{\frac{q}{1-q}}. \tag{2}$$

Moreover, set

$$q' = 2 - \frac{1}{q}, \qquad \sigma'^2 = \frac{3-q}{1+q} \sigma^2.$$

Then the q-escort distribution of $p_q(x; \mu, \sigma)$ is a q-escort distribution, that is, $p_{q'}(x; \mu, \sigma') = P_q(x; \mu, \sigma)$.

This proposition implies that a q-normal distribution generates another q-normal distribution through the escort distribution. This procedure is called the τ-transformation. We remark that the normalization $\Omega_{q,p} = \Omega_q(p)$ of the q-escort distribution is given by

$$\Omega_{q,p} = \frac{3-q}{2} Z_{q,\sigma}^{1-q}. \tag{3}$$

In fact, from Equation (2), we obtain

$$P_q(x; \mu, \sigma) = \frac{1}{\Omega_{q,p}} p_q(x)^q$$

$$= \frac{2}{3-q} \frac{1}{Z_{q,\sigma}} \left[1 - \frac{1-q}{3-q} \frac{(x-\mu)^2}{\sigma^2} \right]_+^{\frac{q}{1-q}}$$

$$= \frac{2}{3-q} \frac{1}{Z_{q,\sigma}} \{ Z_{q,\sigma} p_q(x; \mu, \sigma) \}^q$$

$$= \frac{2}{3-q} Z_{q,\sigma}^{q-1} p_q(x; \mu, \sigma)^q.$$

This implies that the q-escort distribution is given by (2), and the normalization by (3).

4. Geometry of q-normal distributions

In this section, we consider geometry of q-normal distributions. From now on, we suppose that $1 < q < 3$. In the case $q < 1$, the support of a q-normal distribution depends on its parameters, so the q-normal family is not a statistical model in the sense of information geometry.

Suppose that $S_q = \{ p_q(x; \mu, \sigma) \}$ is the set of q-normal distributions. We call S_q the q-*normal family*. In the same arguments as the normal family, we regard S_q as a manifold with a local coordinate $\{ \mu, \sigma \}$. In this case, the local coordinate neighborhood Ξ coincides with the upper half plane.

We denote by g^F the Fisher metric on a q-normal family. We remark that the Fisher metric g^F can be defined on S_q though a q-normal distribution may not have a mean and a variance.

Let us consider Riemannian geometric structures of q-normal families. In order to describe geometry of q-normal families, here we recall the q-exponential and the q-logarithm functions. For a fixed positive number q, the q-*exponential function* is defined by

$$\exp_q x := \begin{cases} (1 + (1-q)x)^{\frac{1}{1-q}}, & q \neq 1, \quad (1 + (1-q)x > 0), \\ \exp x, & q = 1, \end{cases} \tag{4}$$

and the q-*logarithm function* by

$$\ln_q x := \begin{cases} \dfrac{x^{1-q} - 1}{1-q}, & q \neq 1, \ (x > 0), \\ \ln x, & q = 1. \end{cases}$$

Taking a limit $q \to 1$, the standard exponential and the standard logarithm are recovered from the q-exponential and the q-logarithm, respectively.

For a q-normal distribution $p_q(x; \mu, \sigma)$, set

$$\theta^1 = \frac{2}{3-q} Z_{q,\sigma}^{q-1} \cdot \frac{\mu}{\sigma^2} = \frac{2}{3-q} A_q^{q-1} \mu \sigma^{q-3} = \Omega_{q,p}^{-1} \frac{\mu}{\sigma^2},$$

$$\theta^2 = -\frac{1}{3-q} Z_{q,\sigma}^{q-1} \cdot \frac{1}{\sigma^2} = -\frac{1}{3-q} A_q^{q-1} \sigma^{q-3} = -\frac{1}{2} \Omega_{q,p}^{-1} \frac{1}{\sigma^2},$$

then we have

$$\sigma = (3-q)^{\frac{1}{q-3}} A_q^{\frac{q-1}{3-q}} (-\theta^2)^{\frac{1}{q-3}},$$

$$Z_{q,\theta} = A_q \sigma = (3-q)^{\frac{1}{q-3}} A_q^{\frac{2}{3-q}} (-\theta^2)^{\frac{1}{q-3}} = \left\{ \frac{A_q}{\sqrt{(3-q)(-\theta^2)}} \right\}^{\frac{2}{3-q}}.$$

In addition, using the normalization $\Omega_{q,p}$ in (3), set

$$\psi(\theta) = -\frac{(\theta^1)^2}{4\theta^2} - \frac{Z_{q,\sigma}^{q-1} - 1}{1-q}$$

$$= -\frac{(\theta^1)^2}{4\theta^2} + \frac{3-q}{2(q-1)} \Omega_{q,p}^{-1} + \frac{1}{1-q}, \tag{5}$$

$$F_1(x) = x, \quad F_2(x) = x^2,$$

then we obtain

$$\ln_q p_q(x) = \frac{1}{1-q} (p_q(x)^{1-q} - 1)$$

$$= \frac{1}{1-q} \left\{ \frac{1}{Z_{q,\sigma}^{1-q}} \left(1 - \frac{1-q}{3-q} \frac{(x-\mu)^2}{\sigma^2} \right) - 1 \right\}$$

$$= \frac{2\mu Z_{q,\sigma}^{q-1}}{(3-q)\sigma^2} x - \frac{Z_{q,\sigma}^{q-1}}{(3-q)\sigma^2} x^2 - \frac{Z_{q,\sigma}^{q-1}}{3-q} \cdot \frac{\mu^2}{\sigma^2} + \frac{Z_{q,\sigma}^{q-1} - 1}{1-q}$$

$$= \theta^1 F_1(x) + \theta^2 F_2(x) - \psi(\theta).$$

This implies that the q-normal family is a q-exponential family.

From now on, for simplicity, set $\partial_i = \partial/\partial\theta^i$.

Proposition 4.1. *Suppose that S_q is the q-normal family, and ψ is the function defined by Equation (5). Then ψ is strictly convex.*

Proof. Differentiate twice the function ψ, we have the Hessian of ψ as

$$g_{ij}^q(\theta) = \partial_i \partial_j \psi(\theta),$$

$$g^q(\theta) = (g_{ij}^q(\theta)) = -\frac{1}{2\theta^2} \begin{pmatrix} 1 & -\frac{\theta^1}{\theta^2} \\ -\frac{\theta^1}{\theta^2} & \frac{(\theta^1)^2}{(\theta^2)^2} - \frac{2}{3-q} \Omega_{q,p}^{-1} \frac{1}{\theta^2} \end{pmatrix}. \tag{6}$$

Since $1 < q < 3$ and $0 < \sigma < \infty$, from straightforward calculations, we obtain

$$g_{11}^q(\theta) = -\frac{1}{2\theta^2} > 0,$$

$$\det(g_{ij}^q(\theta)) = \frac{1}{2(3-q)}\Omega_{q,p}^{-1}\frac{1}{(-\theta^2)^3} > 0.$$

This implies that ψ is strictly convex.   □

Since g^q is positive definite, it determines a Riemannian metric on S_q. We call g^q the *q-Fisher metric* on S_q. The q-Fisher metric g^q recovers the Fisher metric g^F if q goes to one. Recall that the Fisher metric on the normal family S is a Hessian metric. The q-Fisher metric g^q on S_q is not the Fisher metric, but a Hessian metric.

We define the *q-expectation parameters* (η_1, η_2) by

$$\eta_1 = E_{q,p}[F_1(x)] = \mu,$$
$$\eta_2 = E_{q,p}[F_2(x)] = \mu^2 + \sigma^2.$$

It is known that $\{\eta_i\}$ is the dual coordinate system of $\{\theta^i\}$ with respect to the q-Fisher metric g^q ([10]). The q-Fisher metric in the $\{\eta_i\}$-coordinates and its potential function are given by

$$g^q(\eta) = \frac{1}{(\eta_2 - (\eta_1)^2)^2}\,\Omega_{q,p}^{-1}\begin{pmatrix} \eta_2 + (2-q)(\eta_1)^2 & -\dfrac{3-q}{2}\,\eta_1 \\[2mm] -\dfrac{3-q}{2}\eta_1 & \dfrac{3-q}{4} \end{pmatrix},$$

$$\phi_q(\eta) = -\frac{1}{1-q}\,(\Omega_{q,p}^{-1} - 1).$$

The parameter changes between the $\{\theta^i\}$- and the $\{\eta^i\}$-coordinates are given as follows:

$$\eta_1 = -\frac{\theta^1}{2\theta^2} \qquad\qquad = \frac{\partial}{\partial\theta^1}\psi(\theta),$$

$$\eta_2 = \frac{(\theta^1)^2}{4(\theta^2)^2} - \frac{1}{2\theta^2}\Omega_{q,p}^{-1} \qquad = \frac{\partial}{\partial\theta^2}\psi(\theta),$$

$$\theta^1 = \frac{\eta_1}{\eta_2 - (\eta_1)^2}\Omega_{q,p}^{-1} \qquad\qquad = \frac{\partial}{\partial\eta_1}\phi(\eta),$$

$$\theta^2 = -\frac{1}{2(\eta_2 - (\eta_1)^2)}\Omega_{q,p}^{-1} \qquad = \frac{\partial}{\partial\eta_2}\phi(\eta).$$

We note that the η-potential function $\phi_q(\eta)$ is given by the q-expectation of the q-log likelihood function. In fact, the following holds:

$$E_{q,p}[\ln_q p_q(x;\eta)] = \int_{-\infty}^{\infty} \Omega_{q,p}^{-1} p(x;\eta)^q \cdot \frac{1}{1-q} \left(p(x;\eta)^{1-q} - 1\right) dx$$

$$= \frac{1}{1-q} \Omega_{q,p}^{-1} \int_{-\infty}^{\infty} (p(x;\eta) - p(x;\eta)^q) \, dx$$

$$= \frac{1}{1-q} \left(\Omega_{q,p}^{-1} - 1\right)$$

$$= \phi_q(\eta).$$

Next, let us consider a relation between the standard Fisher metric and the q-Fisher metric.

Proposition 4.2. *Let $S_q = \{p_q(x;\theta)\}$ be the q-normal family, let g^F and g^q be the Fisher metric and the q-Fisher metric on S_q, respectively. Denote by $\Omega_{q,p}$ the normalization of the escort distribution of $p_q(x;\xi)$. Then g^F and g^q satisfy*

$$g^q(\theta) = \frac{q}{\Omega_{q,p}} g^F(\theta),$$

that is, two metrics g^F and g^q are conformally equivalent to each other.

Proof. Set the q-log likelihood function by $l_{q,\theta} = \ln_q p_q(x;\theta)$. From direct calculations, we have

$$\partial_i p_q(x;\theta) = p_q(x;\theta)^q (F_i(x) - \partial_i \psi(\theta)),$$

$$\partial_i \partial_j p_q(x;\theta) = p_q(x;\theta)^{2q-1}(F_i(x) - \partial_i \psi(\theta))(F_j(x) - \partial_j \psi(\theta))$$

$$+ p_q(x;\theta)^q \partial_i \partial_j \psi(\theta), \qquad (7)$$

$$\partial_i \ln p_q(x;\theta) = \frac{\partial_i p_q(x;\theta)}{p_q(x;\theta)} = p_q(x;\theta)^{q-1}(F_i(x) - \partial_i \psi(\theta))$$

$$= p_q(x;\theta)^{q-1}(\partial_i l_{q,\theta}).$$

From the definition of the Fisher metric (1), we have

$$g_{ij}^F(\theta) = E_\theta[(\partial_i \ln p_q(x;\theta))(\partial_j \ln p_q(x;\theta))]$$

$$= \int_{-\infty}^{\infty} p_q(x;\theta)^{2q-2}(\partial_i l_{q,\theta})(\partial_j l_{q,\theta}) p_q(x;\theta) dx$$

$$= \int_{-\infty}^{\infty} p_q(x;\theta)^{2q-1}(\partial_i l_{q,\theta})(\partial_j l_{q,\theta}) dx.$$

Since

$$\int \partial_i p_q(x,\theta) \, dx = \partial_i \int p_q(x,\theta) \, dx = 0,$$

we have

$$\int \partial_i \partial_j p_q(x, \theta) \, dx = 0.$$

Integrating both sides of Equation (7), we obtain

$$g_{ij}^F(\theta) = \frac{1}{q} \int_{-\infty}^{\infty} p_q(x; \theta)^q \, \partial_i \partial_j \psi(\theta) \, dx$$

$$= \frac{\Omega_{q,p}}{q} \, g_{ij}^q(\theta). \qquad \qquad \square$$

From Equation (6), the Fisher metric in the (θ^1, θ^2)-parametrization is given as follows:

$$g^F(\theta) = (g_{ij}^F(\theta)) = -\frac{\Omega_{q,p}}{2q\theta^2} \begin{pmatrix} 1 & -\dfrac{\theta^1}{\theta^2} \\ -\dfrac{\theta^1}{\theta^2} & \dfrac{(\theta^1)^2}{(\theta^2)^2} - \dfrac{2}{3-q} \Omega_{q,p}^{-1} \dfrac{1}{\theta^2} \end{pmatrix}. \qquad (8)$$

Recall that we assumed $1 < q < 3$. Then we obtain the following theorem.

Theorem 4.1. *Let $S_q = \{p_q(x; \theta)\}$ be the q-normal family, and g^F the Fisher metric on S_q. Then the Riemannian manifold (S_q, g^F) is a space of constant negative sectional curvature $\kappa = -q/(3-q)$.*

Proof. From a representatipm of the Fisher metric (8), using a computer, we can directly obtain the result. Here, we show the result by calculating the Christoffel symbols. Since $\partial_1 \Omega_{p,q} = 0$ and

$$\partial_2 \Omega_{q,p} = \frac{(q-1)\Omega_{q,p}}{(3-q)\theta^2},$$

derivatives of the Fisher metric g^F are given by

$$\partial_1 g_{11}^F = 0, \quad \partial_1 g_{12}^F = \frac{\Omega_{q,p}}{2q} \cdot \frac{1}{(\theta^2)^2}, \quad \partial_1 g_{22}^F = -\frac{\Omega_{q,p}}{q} \cdot \frac{\theta^1}{(\theta^2)^3},$$

$$\partial_2 g_{11}^F = \frac{(2-q)\Omega_{q,p}}{q(3-q)} \cdot \frac{1}{(\theta^2)^2}, \quad \partial_2 g_{12}^F = \frac{(3q-7)\Omega_{q,p}}{2q(3-q)} \cdot \frac{\theta^1}{(\theta^2)^3},$$

$$\partial_2 g_{22}^F = \frac{(5-2q)(\theta^1)^2 - 2\theta^2}{q(3-q)(\theta^2)^4}.$$

Hence, Christoffel symbols of the first kind are given by

$$\Gamma^F_{11,1} = 0, \qquad \Gamma^F_{11,2} = \frac{\Omega_{q,p}}{2q(3-q)} \cdot \frac{1}{(\theta^2)^2},$$

$$\Gamma^F_{12,1} = \Gamma^F_{21,1} = \frac{(2-q)\Omega_{q,p}}{2q(3-q)} \cdot \frac{1}{(\theta^2)^2},$$

$$\Gamma^F_{12,2} = \Gamma^F_{21,2} = -\frac{\Omega_{q,p}}{2q} \cdot \frac{\theta^1}{(\theta^2)^3},$$

$$\Gamma^F_{22,1} = \frac{(q-2)\Omega_{q,p}}{q(3-q)} \cdot \frac{\theta^1}{(\theta^2)^3}, \quad \Gamma^F_{22,2} = \frac{(5-2q)\Omega_{q,p}(\theta^1)^2 - 2\theta^2}{2q(3-q)(\theta^2)^4}.$$

The Riemannian curvature tensor of g^F is determined by

$$R^1_{212} = \partial_1\Gamma^F_{22}{}^1 - \partial_2\Gamma^F_{12}{}^1 + \Gamma^F_{11}{}^1\Gamma^F_{22}{}^1 + \Gamma^F_{12}{}^1\Gamma^F_{22}{}^2 - \Gamma^F_{21}{}^1\Gamma^F_{12}{}^1 - \Gamma^F_{22}{}^1\Gamma^F_{22}{}^2$$

$$= \frac{(3-q)\Omega_{q,p}(\theta^1)^2 + 2\theta^2}{2(3-q)^2(\xi^2)^3},$$

$$R^2_{212} = \frac{\Omega_{q,p}\theta^1}{2(3-q)(\theta^2)^2}.$$

Hence we obtain

$$R_{1212} = g_{11}R^1_{212} + g_{12}R^2_{212} = \frac{\Omega_{q,p}}{2q(3-q)^2} \cdot \frac{1}{(\theta^2)^3}.$$

From Equation (8), the determinant of the Fisher metric is given by

$$\det(g^F_{ij}) = -\frac{\Omega_{q,p}}{2q^2(3-q)} \cdot \frac{1}{(\theta^2)^3}.$$

Therefore, we finally obtain the sectional curvature as

$$\kappa = \frac{R_{1212}}{\det(g^F_{ij})} = -\frac{q}{3-q}. \qquad \square$$

We note that the sectional curvature κ goes to $-1/2$ in the case $q \to 1$.

In information geometry, a pair of dual affine connections is usually discussed [1]. However, we did not mention such dual affine connections in this paper. Geometry of dual affine connections is discussed in [5] for more generalized probability distributions. For the set of discrete probability distributions, geometry of the q-Fisher metric and the q-escort distributions is studied in [6].

Acknowledgment

The authors wish to express their sincere gratitude to the referee for his appropriate comments of the paper.

The third named author is partially supported by The Toyota Physical and Chemical Research Institute and by Grant-in-Aid for Encouragement of Young Scientists (B) No. 19740033, Japan Society for the Promotion of Science.

References

1. S. Amari and H. Nagaoka, *Methods of information geometry*, Amer. Math. Soc., Providence, Oxford University Press, Oxford, 2000.
2. C.M. Bishop, *Pattern Recognition And Machine Learning*, Springer, 2006.
3. S.L. Lauritzen, *Statistical manifolds*, Differential Geometry in Statistical Inferences, IMS Lecture Notes Monograph Series 10, Institute of Mathematical Statistics, Hayward California, (1987), 96–163.
4. J. Naudts, *Generalised Thermostatistics*, Springer, 2011.
5. H. Matsuzoe and A. Ohara, *Geometry for q-exponential families*, Recent Progress in Differential Geometry and its Related Fields, Proceedings of the 2nd International Colloquium on Differential Geometry and its Related Fields, World Scientific, (2011), 55–71.
6. A. Ohara, H. Matsuzoe and S. Amari, *A dually flat structure on the space of escort distributions*, J. Phys.: Conf. Ser. **201**(2010), No. 012012 (electronic).
7. H. Shima, *The Geometry of Hessian Structures*, World Scientific, 2007.
8. H. Suyari and M. Tsukada, *Law of Error in Tsallis Statistics*, IEEE Trans. Inform. Theory, **51**(2005), 753–757.
9. M. Tanaka, *A Consideration on a Family of q-Normal Distributions*, J. Inst. Elect. Inf. Comm. Eng. Japan, **J85-D-II**(2)(2002), 161–173. (Japanese)
10. M. Tanaka, *Geometry of non-exponential family and α-connection through Gauge theory*, preprint. (Japanese)
11. M. Tanaka, *Meaning of an escort distribution and τ-transformation*, J. Phys.: Conf. Ser. **201**(2010), No 012007 (electronic).
12. C. Tsallis, *Introduction to Nonextensive Statistical Mechanics: Approaching a Complex World*, Springer, New York, 2009.

Received April 24, 2011
Revised May 3, 2011

Proceedings of the 2nd International
Colloquium on Differential Geometry
and its Related Fields
Veliko Tarnovo, September 6–10, 2010

A REMARK ON COMPLEX LAGRANGIAN CONES IN $\mathbf{H}^n$

Norio EJIRI

Department of Mathematics, Meijo University,
Shiogamaguti, 1-501, Tenpaku-ku,
Nagoya 468-8502, Japan
E-mail: ejiri@meijo-u.ac.jp

Kazumi TSUKADA

Department of Mathematics, Ochanomizu University,
Otsuka, Bunkyou-ku,
Tokyo 112-8610, Japan
E-mail: tsukada@math.ocha.ac.jp

A complex Lagrangian cone in $\mathbf{H}^n$ gives an integral submanifold (holomorphic horizontal submanifold) and a minimal Lagrangian submanifold in a complex projective space. We clarify the relationship between two submanifolds.

Keywords: Complex Lagrangian cone, Complex symplectic structure, Holomorphic contact structure, Complex projective space.

1. Introduction

Let $\mathbf{H}$ be the real division algebra of quaternions generated by $1, i, j, k$ and $\mathbf{H}^n$ be the quaternionic vector space with canonical Euclidean inner product $\langle\ ,\ \rangle$. Each element of a basis $\{i, j, k\}$ of purely imaginary quaternions induces an orthogonal complex structure corresponding to the right action. We may see that $\mathbf{H}^n$ is a $2n$-dimensional complex Euclidean vector space with respect to i, j, k. Let N be a real $2n$-dimensional submanifold of $\mathbf{H}^n$. We call N *complex Lagrangian* if either N is Lagrangian with respect to i and is complex with respect to j or it is complex with respect to i and is Lagrangian with respect to j. Since $k = ij$, a Lagrangian submanifold N is again Lagrangian with respect to k. Holomorphic curves in $(\mathbf{C}^2, i)$ are complex Lagrangian and special Lagrangian surfaces in $(\mathbf{C}^2, i)$. They are complex with respect to the orthogonal complex structure j, hence they are complex Lagrangian (see [5]). Ionel [4] proved local existence theorems (Theorem 3.24 and Theorem 3.25) on the complex Lagrangian submanifold in $\mathbf{C}^4$ by use of the Cartan-Kähler theorem. When we consider $\mathbf{H}^n = T^*\mathbf{C}^n$, the complex Lagrangian graph of a holomorphic function on a domain of $\mathbf{C}^n$ is a complex Lagrangian submanifold. If N includes the origin o of $\mathbf{C}^n$ and satisfies $tN = N$ for all $t \in \mathbf{R}$, then it is called a *complex Lagrangian cone*. Let $(\mathbf{C}P^{2n-1}, i), (\mathbf{C}P^{2n-1}, j)$ be complex projective spaces of holomorphic sectional curvature 4 whose complex structures are defined by i, j, respectively. Then a complex Lagrangian cone N gives a minimal Lagrangian submanifold in $(\mathbf{C}P^{2n-1}, i)$ and a complex submanifold in $(\mathbf{C}P^{2n-1}, j)$. We note that the odd dimensional complex projective space $\mathbf{C}P^{2n-1}$ admits a holomorphic contact structure induced from $\mathbf{C}^{2n}$. We call a maximal integral submanifold of $\mathbf{C}P^{2n-1}$ (with respect to the holomorphic contact structure) an *integral manifold* (holomorphic horizontal submanifold). In order to make clear the complex structure on $\mathbf{C}P^{2n-1}$, we denote fibrations in the following manner: $\pi : \mathbf{C}^{2n} \setminus \{0\} \to \mathbf{C}P^{2n-1}$, $\pi_j : S^{4n-1} \to (\mathbf{C}P^{2n-1}, j)$ and $\pi_i : S^{4n-1} \to (\mathbf{C}P^{2n-1}, i)$.

Theorem 1.1. *Let M_j be an $(n-1)$-dimensional complex manifold in $(\mathbf{C}P^{2n-1}, j)$. This manifold M_j is the integral manifold in $(\mathbf{C}P^{2n-1}, j)$ with respect to the holomorphic contact structure if and only if the cone $\pi_j^{-1}(M_j)$ is complex Lagrangian (complex with respect to j and Lagrangian with respect to i). The submanifold $\pi_i(\pi^{-1}(M_j))$ is minimal Lagrangian in $(\mathbf{C}P^{2n-1}, i)$ which is considered as the set of equators of the Penrose fibration at a point (as north pole) of M_j.*

In [3], we give a point of view of Lagrangian submanifolds in $\mathbf{C}P^{2n-1}$. Our theorem gives another proof of the theorem in [1] and the explicit relationship between integral manifolds and their minimal Lagrangian submanifolds in $\mathbf{C}P^{2n-1}$ by use of complex Lagrangian cone in $\mathbf{C}^{2n}$. Our theorem is a generalization of the result in [1].

2. A proof of Theorem

Let $\mathbf{H}^n$ be the space of column n-tuples with entries in the field $\mathbf{H}$. The space $\mathbf{H}^n$ is considered as a right $\mathbf{H}$-vector space, i.e., vectors are multiplied by quaternions from the right. It can be considered as a complex vector space in a variety of natural ways. We choose a real linear transformation as a complex structure and define a complex scalar multiplication from the right on $\mathbf{H}^n$ by $\boldsymbol{x}(a + bi) = a\boldsymbol{x} + b\boldsymbol{x}i$ for $a, b \in \mathbf{R}$ and $\boldsymbol{x} \in \mathbf{H}^n$. Let $\{e_1, e_2, \cdots, e_n\}$ be the standard basis of quaternionic vector space $\mathbf{H}^n$. That is, e_i is the vector of $\mathbf{H}^n$ whose i-th component is 1 and the other components are zero. Then $\{e_1, e_2, \cdots, e_n, e_1j, e_2j, \cdots, e_nj\}$ is a complex basis of $\mathbf{H}^n$. Using this complex basis, we identify $\mathbf{H}^n$ with $\mathbf{C}^{2n}$: For

$\boldsymbol{x} = {}^t(x_1, x_2, \cdots, x_n) \in \mathbf{H}^n$, we take $z_\alpha, z_{n+\alpha} \in \mathbf{C}$ so that $x_\alpha = z_\alpha + j z_{n+\alpha}$ ($\alpha = 1, 2, \cdots, n$). We can identtify $\mathbf{H}^n$ with $\mathbf{C}^{2n}$ by

$$\mathbf{H}^n \ni \boldsymbol{x} \mapsto \begin{pmatrix} z_1 \\ \vdots \\ z_n \\ z_{n+1} \\ \vdots \\ z_{2n} \end{pmatrix} \in \mathbf{C}^{2n}.$$

From now on we write $\mathbf{C}^{2n}$ for $\mathbf{H}^n$. We define a $\mathbf{C}$-Hermitian inner product $\langle \, , \, \rangle_{\mathbf{C}}$ and a $\mathbf{C}$-skew symmetric form Ω on $\mathbf{C}^{2n}$ as follows:

$$\langle \boldsymbol{z}, \boldsymbol{w} \rangle_{\mathbf{C}} = {}^t \bar{\boldsymbol{z}} \boldsymbol{w} = \sum_{\alpha=1}^{2n} \bar{z}_\alpha w_\alpha,$$

$$\Omega(\boldsymbol{z}, \boldsymbol{w}) = \sum_{\alpha=1}^{n} (z_\alpha w_{n+\alpha} - z_{n+\alpha} w_\alpha),$$

for any $\boldsymbol{z}, \boldsymbol{w} \in \mathbf{C}^{2n}$. Then we have

$$\langle \boldsymbol{z}j, \boldsymbol{w} \rangle_{\mathbf{C}} = \Omega(\boldsymbol{z}, \boldsymbol{w}). \tag{1}$$

Hence the $\mathbf{C}$-skew symmetric form Ω is the Kähler form for j. We define a holomorphic 1-form ω on $\mathbf{C}^{2n}$ by

$$\omega = \frac{1}{2} \sum_{\alpha=1}^{n} (z_\alpha dz_{n+\alpha} - z_{n+\alpha} dz_\alpha).$$

Then it follows that

$$d\omega = \sum_{\alpha=1}^{n} dz_\alpha \wedge dz_{n+\alpha}.$$

We can identify $d\omega$ with the skew symmetric form Ω by the usual identification of the tangent space $T_z \mathbf{C}^{2n}$ with $\mathbf{C}^{2n}$. This form is called a complex symplectic form. Let η be a radial vector field defined by $\eta_z = \sum_{\alpha=1}^{2n} z_\alpha \dfrac{\partial}{\partial z_\alpha}$ at $z = {}^t(z_1, \cdots, z_{2n}) \in \mathbf{C}^{2n}$. Then we have $\omega(\eta) = 0$ and $\iota_\eta d\omega = d\omega(\eta, \cdot) = 2\omega$. Let M^n be an n-dimensional complex submanifold of $\mathbf{C}^{2n}$. By (2.1), one can easily see that M^n is complex Lagrangian if and only if $\Omega = d\omega$ vanishes on M^n, that is, $\Omega(X, Y) = 0$ for $X, Y \in T_z M^n$ $z \in M^n$. Let $\pi : \mathbf{C}^{2n} \setminus \{0\} \to \mathbf{C}P^{2n-1}$ be a natural projection, which is a principal fibre bundle with the structure group $\mathbf{C}^* = \mathbf{C} \setminus \{0\}$. We define a holo-morphic contact structure $\mathcal{H}$ on $\mathbf{C}P^{2n-1}$ as follows: Let $\{U_\lambda\}$ be an open covering of $\mathbf{C}P^{2n-1}$ and s_λ be a holomorphic cross section of the princi-pal fibre bundle $\pi : \mathbf{C}^{2n} \setminus \{0\} \to \mathbf{C}P^{2n-1}$ over U_λ. Set $\omega_\lambda = s_\lambda^* \omega$. On $U_\lambda \cap U_\mu$, there exists a non-vanishing holomorphic function $f_{\lambda\mu}$ such that $s_\lambda = f_{\lambda\mu} s_\mu$. Then $\omega_\lambda = f_{\lambda\mu}^2 \omega_\mu$ on $U_\lambda \cap U_\mu$. Therefore $\omega_\lambda = 0$ defines a $2(n-1)$-dimensional subbundle $\mathcal{H}$ of $T\mathbf{C}P^{2n-1}$. Moreover on $\mathcal{H}$, the skew-symmetric form $d\omega_\lambda = s_\lambda^* d\omega = s_\lambda^* \Omega$ is non-degenerate. We call $\mathcal{H}$ a *holomorphic contact structure*.

Proposition 2.1. *Let M be an $(n-1)$-dimensional complex submanifold in $\mathbf{C}P^{2n-1}$ which is an integral manifold of $\mathcal{H}$. The pull-back $\pi^{-1}(M)$ in $\mathbf{C}^{2n} \setminus \{0\}$ is an n-dimensional complex Lagrangian cone. Conversely any n-dimensional complex Lagrangian cone N in $\mathbf{C}^{2n} \setminus \{0\}$ defines an $(n-1)$-dimensional complex submanifold $\pi(N)$ in $\mathbf{C}P^{2n-1}$ which is an integral manifold of $\mathcal{H}$.*

Proof. *The former part of Theorem.* We will prove that the form $\Omega = d\omega$ vanishes on $\pi^{-1}(M)$. For an arbitrary point $z \in \pi^{-1}(M)$, we put $x = \pi(z) \in M$. We take a neighborhood U of x in $\mathbf{C}P^{2n-1}$ and a holomorphic cross section s over U with $s(x) = z$. Then the tangent space $T_z\pi^{-1}(M)$ is spanned by η_z and s_*T_xM. Since $s^*d\omega = ds^*\omega = 0$ on M, we have $d\omega(s_*X, s_*Y) = 0$ for $X, Y \in T_xM$. Moreover we have $d\omega(\eta_z, s_*X) = 2\omega(s_*X) = 2s^*\omega(X) = 0$ for $X \in T_xM$. These imply that $\Omega = d\omega$ vanishes on $\pi^{-1}(M)$.

The latter part of Theorem. Since N is a cone, we have $\eta_z \in T_zN$ at any point $z \in N$. Therefore $\pi(N)$ is an $(n-1)$-dimensional complex submanifold in $\mathbf{C}P^{2n-1}$, which is denoted by M. For any tangent vector $X \in T_{\pi(z)}M$, we take a tangent vector $\tilde{X} \in T_zN$ such that $\pi_*(\tilde{X}) = X$. Since N is complex Lagrangian, we have $0 = d\omega(\eta_z, \tilde{X}) = 2\omega(\tilde{X})$. This implies that M is an integral submanifold of $\mathcal{H}$. $\qquad\square$

Let N be a complex Lagrangian cone which is Lagrangian with respect to i and complex with respect to j. Then there is a $(2n-1)$-dimensional submanifold M in $S^{4n-1}(1)$ such that $N = \{tp \in \mathbf{H}^n : t > 0, p \in M\} \cup \{0\}$. Let π_j be the Hopf map of $S^{4n-1}(1)$ onto $(\mathbf{C}P^{2n-1}, j)$, where $(\mathbf{C}P^{2n-1}, j)$ is the complex projective space with complex structutre j. There is a hor-izontal complex submanifold M_j in $(\mathbf{C}P^{2n-1}, j)$ satisfying $M = \pi_j^{-1}(M_j)$. On the other hand, since M is minimal Legendrian, M is isometrically im-mersed in $(\mathbf{C}P^{2n-1}, i)$ as a Lagrangian submanifold M'. Thus we obtain two minimal submanifolds M_j in $(\mathbf{C}P^{2n-1}, j)$ and M' in $(\mathbf{C}P^{2n-1}, i)$. We obtain the following.

Proposition 2.2. *The above submanifold M_j in $(\mathbf{C}P^{2n-1}, j)$ is identified with the lift to the twisor space $(\mathbf{C}P^{2n-1}, j)$ of the totally complex submanifold $\pi(M)$ in $\mathbf{H}P^{n-1}(4)$ where π is the projection of $S^{4n-1}(1)$ onto $\mathbf{H}P^{n-1}(4)$.*

Proof. Since $M = \pi_j^{-1}(M_j)$, it is a circle bundle over M_j. Geodesic circles of length 2π as fibres of $M \to M_j$ are mapped to geodesic circles of length π in $(\mathbf{C}P^{2n-1}, i)$. Let π be the projection of $S^{4n-1}(1)$ onto $\mathbf{H}P^{n-1}(4)$. Then each fibre $(\simeq S^3(1))$ is identified with an orbit of some point $a \in S^{4n-1}(1)$ by the right action of the unit quaternions. Let η_i be the projection of $(\mathbf{C}P^{2n-1}, i)$ to $\mathbf{H}P^{n-1}(4)$. By the definition of η_i, we see $(\mathbf{C}P^{2n-1}, i)$ is a twistor space of $\mathbf{H}P^{n-1}(4)$ with fibres $S^2(4)$. Similarly, we know that $(\mathbf{C}P^{2n-1}, j)$ is a twistor space of $\mathbf{H}P^{n-1}(4)$ with the projection η_j. A point $a \in M$ can be regard as the position vector of its point in $S^{4n-1}(1)$. We obtain 3 unit vectors ai, aj, ak orthogonal to a such as ai is normal to $T_a N$, aj is tangent to $T_a N$ and hence ak is normal to $T_a N$. Since N is Lagrangian with respect to i and complex with respect to j, we obtain $e_1, e_1 j, ..., e_{n-1}, e_{n-1} j, aj$ as an orthonormal basis of $T_a M$ and $e_1 i, -e_1 k, ..., e_{n-1} i, -e_{n-1} k, ai, ak$ as an orthonormal basis of $T_a^\perp M$, where $T_a^\perp M$ is the normal space of $T_a M$ in $T_a S^{4n-1}(1)$.

Note that

$$\pi_{j*}(e_1), \ \pi_{j*}(e_1 j), \ldots, \pi_{j*}(e_{n-1}), \ \pi_{j*}(e_{n-1} j)$$

is an orthonormal basis of $T_{\pi_j(a)} M_j$,

$$\pi_{j*}(e_1 i), \ \pi_{j*}(-e_1 k), \ldots, \pi_{j*}(e_{n-1} i), \ \pi_{j*}(-e_{n-1} k), \pi_{j*}(ai), \pi_{j*}(ak)$$

is an orthonormal basis of $T_{\pi_j(a)}^\perp M_j$. Since $\pi_{j*}(ai), \pi_{j*}(ak)$ is an orthonormal basis of the tangent space of the fibre at $\pi_{j*}(a)$ of the twistor space $(\mathbf{C}P^{2n-1}, j)$, we find that M_j is the integrable submanifold in the twistor spce $(\mathbf{C}P^{2n-1}, j)$.

Furthermore, since $\eta_j \circ \pi_j = \pi$, we have

$$\pi_*(e_1), \ \pi_*(e_1 j), \ldots, \pi_*(e_{n-1}), \ \pi_*(e_{n-1} j),$$
$$\pi_*(e_1 i), \ \pi_*(-e_1 k), \ldots, \pi_*(e_{n-1} i), \ \pi_*(-e_{n-1} k)$$

is an orthonormal basis of $T_{\pi(a)} \mathbf{H}P^{n-1}$. Thus $\eta_j(M_j) = \pi(M)$ is totally complex in $\mathbf{H}P^{n-1}$ and M_j is its twistor lift. $\qquad\square$

For a complex Lagrangian cone of $\mathbf{H}^n$, we obtain an integrable submanifold M_j in $(\mathbf{C}P^{2n-1}, j)$ and a minimal Lagrangian submanifold M' in $(\mathbf{C}P^{2n-1}, i)$. We shall prove the following.

Proposition 2.3. *The minimal Lagrangian submanifold M' is the circle bundle of a horizontal holomorphic submanifold in $(\mathbf{C}P^{2n-1}, i)$.*

We consider a cone given by $\frac{1}{\sqrt{2}}N(1 + k)$ in $\mathbf{H}^n$. Then we obtain the following.

Lemma 2.1. *The cone $\frac{1}{\sqrt{2}}N(1 + k)$ is complex Lagrangian.*

Proof. As was done in the proof of Proposition2.2 we choose an orthonormal basis such that

$$e_1, e_1 j, ..., e_{n-1}, e_{n-1} j, a, aj$$

is an orthonormal basis of $T_a N$ and $e_1 i, -e_1 k, ..., e_{n-1} i, -e_{n-1} k, ai, ak$ is an orthonormal basis of $T_a^\perp N$, where $T_a^\perp M$ is the normal space of $T_a M$ in $T_a S^{4n-1}(1)$. Then

$$\frac{1}{\sqrt{2}}a(1 + k), \ \frac{1}{\sqrt{2}}aj(1 + k), \ \frac{1}{\sqrt{2}}e_1(1 + k), \ \frac{1}{\sqrt{2}}e_1 j(1 + k), \cdots,$$

$$\frac{1}{\sqrt{2}}e_{n-1}(1 + k), \ \frac{1}{\sqrt{2}}e_{n-1} j(1 + k)$$

is a tangent orthonormal basis at $\frac{1}{\sqrt{2}}a(1 + k)$ of $\frac{1}{\sqrt{2}}N(1 + k)$ and

$$\frac{1}{\sqrt{2}}ai(1 + k), \ \frac{1}{\sqrt{2}}ak(1 + k), \ \frac{1}{\sqrt{2}}e_1 i(1 + k), \ \frac{1}{\sqrt{2}}e_1 k(1 + k), \ \cdots,$$

$$\frac{1}{\sqrt{2}}e_{n-1} i(1 + k), \ \frac{1}{\sqrt{2}}e_{n-1} k(1 + k),$$

is a normal orthonormal basis at $\frac{1}{\sqrt{2}}a(1 + k)$ of $\frac{1}{\sqrt{2}}N(1 + k)$. Since we have

$$(1 + k)i = j(1 + k), \ (1 + k)j = -i(1 + k), \ (1 + k)k = k(1 + k),$$

we see

$$\frac{1}{\sqrt{2}}a(1 + k)i = \frac{1}{\sqrt{2}}aj(1 + k), \ \frac{1}{\sqrt{2}}e_1(1 + k)i = \frac{1}{\sqrt{2}}e_1 j(1 + k), \ \cdots,$$

$$\frac{1}{\sqrt{2}}e_{n-1}(1 + k)i = \frac{1}{\sqrt{2}}e_{n-1} j(1 + k)$$

and

$$\frac{1}{\sqrt{2}}ai(1+k)i = \frac{1}{\sqrt{2}}ak(1+k), \ \frac{1}{\sqrt{2}}e_1i(1+k)i = \frac{1}{\sqrt{2}}e_1k(1+k), \ \cdots,$$
$$\frac{1}{\sqrt{2}}e_{n-1}i(1+k)i = \frac{1}{\sqrt{2}}e_{n-1}k(1+k).$$

These imply that $\frac{1}{\sqrt{2}}N(1+k)$ is complex with respect to i. Similarly we see that $\frac{1}{\sqrt{2}}N(1+k)$ is Lagrangian with respect to j.   □

Proof of Proposition 2.3. By Lemma 2.1 we find $\pi_i(\frac{1}{\sqrt{2}}N(1+k))$ is an integral manifold in $(\mathbf{C}P^{2n-1}, i)$. Note that $\pi(\frac{1}{\sqrt{2}}N(1+k)) = \pi(N)$. We prove the wanted integral manifold.

The geodesic through a in M which is to tangent to aj is mapped to a geodesic circle of the fibre $S^2(4)$ on $\pi(M)$. Note that the geodesic $\cos\theta a + \sin\theta ak$ through a which is tangent to ak is mapped to a geodesic whose tangent is normal to the geodesic circle at $\pi_i(a)$. When $\theta = \pi/4$, we see that $\pi(\frac{1}{\sqrt{2}}a(1+k))$ is a pole for the geodesic circle.   □

Propositions 2.1, 2.2 and 2.3 imply Theorem 1.1.

References

1. N. Ejiri, *Calabi lifting and surface geometry in S^4*, Tokyo J. Math. 9(1986), 297–324.
2. N. Ejiri, *A generating function of a complex Lagrangian cone in* **H**^{n}, preprint.
3. N. Ejiri and K. Tsukada, *Another natural lift of a Kaehler submanifold of a quaternionic Kaehler manifold to the twistor space*, Tokyo J. Math. 28(2005), 71–78.
4. M. Ionel, *Second order families of special Lagrangian submanifold in* **C**^{4}, J. Diff. Geom. 65(2003), 211–272.
5. D. Joyce, *Riemannian holonomy groups and calibrated geometry*, Oxford University Press, 2007.
6. K. Tsukada, *Parallel submanifolds in a quaternion projective space*, Osaka J. Math. 22(1985), 187–241.

Received January 11, 2011
Revised   May 10, 2011

Proceedings of the 2nd International
Colloquium on Differential Geometry
and its Related Fields
Veliko Tarnovo, September 6–10, 2010

REALIZATIONS OF SUBGROUPS OF G_2, $Spin(7)$
AND THEIR APPLICATIONS

Hideya HASHIMOTO* and Misa OHASHI**

Department of Mathematics, Meijo University,
Nagoya 468-8502, Japan
*E-mail: hhashi@meijo-u.ac.jp
**E-mail: m0851501@ccalumni.meijo-u.ac.jp

We concretely describe subgroups of the Lie group $Spin(7)$ and related homogeneous spaces, and give their applications.

Keywords: Octonions, $Spin(7)$, G_2.

1. Introduction

The purpose of this paper is to describe subgroups of the Lie group $Spin(7)$ which is a universal covering space of $SO(7)$. In their paper [3], R.Harvey and H.B.Lawson construct the description of $Spin(7)$. Moreover, R.L.Bryant [1] gave a more concrete representation of $Spin(7)$ by taking account of the algebraic properties of the octonions $\mathfrak{C}$. From these, we can obtain a corresponding homogeneous spaces related to given subgroups. It is known ([3]) that G_2, $SU(4)$, and $Sp(1) \times Sp(1) \times Sp(1)/Z_2$ are subgroups of $Spin(7)$. These subgroups are obtained by the isotropy subgroups of some homogeneous spaces. For example, the unit 7-sphere S^7 can be considered as the homogeneous space $Spin(7)/G_2$. Therefore G_2 is the isotropy subgroup at $1 \in S^7 \subset \mathfrak{C}$. We will write down such relations and fibre bundle structures of homogeneous spaces as explicitly as possible.

2. Preliminaries

Let $\boldsymbol{H}$ be the skew field of all quaternions with canonical basis $\{1, i, j, k\}$ which satisfies

$$i^2 = j^2 = k^2 = -1, \ ij = -ji = k, \ jk = -kj = i, \ ki = -ik = j.$$

The octonions (or Cayley algebra) $\mathfrak{C}$ over $\boldsymbol{R}$ can be considered as a direct sum $\mathbf{H} \oplus \mathbf{H} = \mathfrak{C}$ with the following multiplication

$$(a + b\varepsilon)(c + d\varepsilon) = ac - \bar{d}b + (da + b\bar{c})\varepsilon,$$

where $\varepsilon = (0, 1) \in \mathbf{H} \oplus \mathbf{H}$ and $a, b, c, d \in \mathbf{H}$. Here the symbol " $\bar{}$ " denotes the conjugation of the quaternion. For any $x, y \in \mathfrak{C}$, we have

$$\langle xy, xy \rangle = \langle x, x \rangle \langle y, y \rangle.$$

The pair of such inner product $\langle\ ,\ \rangle$ and product is called "normed algebra" in ([3]). The octonions is a non-commutative, non-associative alternative division algebra. The group of automorphisms of the octonions is the exceptional simple Lie group

$$G_2 = \{g \in SO(8) \mid g(uv) = g(u)g(v) \text{ for any } u, v \in \mathfrak{C}\},$$

where $SO(8)$ is the special orthogonal group of degree 8. In this paper, we shall concern the Lie subgroup $Spin(7)$ which is defined by

$$Spin(7) = \{g \in SO(8) \mid g(uv) = g(u)\chi_g(v) \text{ for any } u, v \in \mathfrak{C}\},$$

where $\chi_g(v) = g(g^{-1}(1)v)$. Note that G_2 is a Lie subgroup of $Spin(7)$:

$$G_2 = \{g \in Spin(7) \mid g(1) = 1\}.$$

The map χ defines a double covering map from $Spin(7)$ onto $SO(7)$ as $\chi(g) = \chi_g$. It satisfies the following equivariance

$$g(u) \times g(v) = \chi_g(u \times v), \tag{1}$$

for any $u, v \in \mathfrak{C}$. Here we set the "exterior product" as $u \times v = (1/2)(\bar{v}u - \bar{u}v)$ with the conjugation $\bar{v} = 2\langle v, 1 \rangle - v$. We note that $u \times v$ is pure-imaginary for any $u, v \in \mathfrak{C}$. In order to describe the structure equations of $Spin(7)$, we fix a basis of the complexification of the octonions $\mathbf{C} \otimes_{\mathbf{R}} \mathfrak{C}$ over $\mathbf{C}$ by

$$N = (1/2)(1 - \sqrt{-1}\varepsilon), \quad \bar{N} = (1/2)(1 + \sqrt{-1}\varepsilon),$$
$$E_1 = iN, \ E_2 = jN, \ E_3 = -kN, \ \bar{E}_1 = i\bar{N}, \ \bar{E}_2 = j\bar{N}, \ \bar{E}_3 = -k\bar{N}.$$

We extend the multiplication of the octonions complex linearly on $\mathbf{C} \otimes_{\mathbf{R}} \mathfrak{C}$. The multiplication table of the octonions is given by

$A\backslash B$	N	E_1	E_2	E_3	$\bar{N}$	$\bar{E}_1$	$\bar{E}_2$	$\bar{E}_3$
N	N	0	0	0	0	$\bar{E}_1$	$\bar{E}_2$	$\bar{E}_3$
E_1	E_1	0	$-\bar{E}_3$	$\bar{E}_2$	0	$-\bar{N}$	0	0
E_2	E_2	$\bar{E}_3$	0	$-\bar{E}_1$	0	0	$-\bar{N}$	0
E_3	E_3	$-\bar{E}_2$	$\bar{E}_1$	0	0	0	0	$-\bar{N}$
$\bar{N}$	0	E_1	E_2	E_3	$\bar{N}$	0	0	0
$\bar{E}_1$	0	$-N$	0	0	$\bar{E}_1$	0	$-E_3$	E_2
$\bar{E}_2$	0	0	$-N$	0	$\bar{E}_2$	E_3	0	$-E_1$
$\bar{E}_3$	0	0	0	$-N$	$\bar{E}_3$	$-E_2$	E_1	0

We also give the table of exterior product $A \times B$.

$A\backslash B$	N	E_1	E_2	E_3	$\bar{N}$	$\bar{E}_1$	$\bar{E}_2$	$\bar{E}_3$
N	0	$-E_1$	$-E_2$	$-E_3$	$-\frac{\sqrt{-1}}{2}\varepsilon$	0	0	0
E_1	E_1	0	$-\bar{E}_3$	$\bar{E}_2$	0	$-\frac{\sqrt{-1}}{2}\varepsilon$	0	0
E_2	E_2	$\bar{E}_3$	0	$-\bar{E}_1$	0	0	$-\frac{\sqrt{-1}}{2}\varepsilon$	0
E_3	E_3	$-\bar{E}_2$	$\bar{E}_1$	0	0	0	0	$-\frac{\sqrt{-1}}{2}\varepsilon$
$\bar{N}$	$\frac{\sqrt{-1}}{2}\varepsilon$	0	0	0	0	$-\bar{E}_1$	$-\bar{E}_2$	$-\bar{E}_3$
$\bar{E}_1$	0	$\frac{\sqrt{-1}}{2}\varepsilon$	0	0	$\bar{E}_1$	0	$-E_3$	E_2
$\bar{E}_2$	0	0	$\frac{\sqrt{-1}}{2}\varepsilon$	0	$\bar{E}_2$	E_3	0	$-E_1$
$\bar{E}_3$	0	0	0	$\frac{\sqrt{-1}}{2}\varepsilon$	$\bar{E}_3$	$-E_2$	E_1	0

We define the $Spin(7)$ frame field $(n, f, \bar{n}, \bar{f})$ as follows;

$$(n, f, \bar{n}, \bar{f}) = (g(N), g(E), g(\bar{N}), g(\bar{E})) = (N, E, \bar{N}, \bar{E}) \begin{pmatrix} P & \bar{Q} \\ Q & \bar{P} \end{pmatrix},$$

where $g \in Spin(7)$ and $\begin{pmatrix} P & \bar{Q} \\ Q & \bar{P} \end{pmatrix}$ is an its matrix representation. Here P is a $SU(4)$-valued function and Q is a 4×4 matrix-valued function on $Spin(7)$. The Lie algebra of this representaion is given by

Proposition 2.1 ([1]). *The Maurer-Cartan form of $Spin(7)$ is given by*

$$d(n, f, \bar{n}, \bar{f}) = (n, f, \bar{n}, \bar{f}) \left(\begin{array}{cc|cc} \sqrt{-1}\rho & -{}^t\bar{\mathfrak{h}} & 0 & -{}^t\theta \\ \mathfrak{h} & \kappa & \theta & [\bar{\theta}] \\ \hline 0 & -{}^t\bar{\theta} & -\sqrt{-1}\rho & -{}^t\mathfrak{h} \\ \bar{\theta} & [\theta] & \mathfrak{h} & \bar{\kappa} \end{array} \right) \tag{2}$$

$$= (n, f, \bar{n}, \bar{f})\psi,$$

where ψ is the $spin(7)(\subset M_{8\times 8}(\mathbf{C}))$-valued 1-form, ρ is a real-valued 1-form, $\mathfrak{h}, \theta$ are $M_{3\times 1}$-valued 1-form, κ is a $\mathfrak{u}(3)$-valued 1-form which satisfy $\sqrt{-1}\rho + tr\kappa = 0$, and

$$[\theta] = \begin{pmatrix} 0 & \theta^3 & -\theta^2 \\ -\theta^3 & 0 & \theta^1 \\ \theta^2 & -\theta^1 & 0 \end{pmatrix}$$

for $\theta = {}^t(\theta^1, \theta^2, \theta^3)$. The ψ satisfy the following integrability condition $d\psi + \psi \wedge \psi = 0$. More precisely

$$dn = n\sqrt{-1}\rho + f\mathfrak{h} + \bar{f}\bar{\theta},$$
$$df = n(-{}^t\bar{\mathfrak{h}}) + f\kappa + n(-{}^t\bar{\theta}) + \bar{f}[\theta],$$

and the integrability conditions are given by

$$d(\sqrt{-1}\rho) = {}^t\bar{\mathfrak{h}} \wedge \mathfrak{h} + {}^t\theta \wedge \bar{\theta},$$
$$d\mathfrak{h} = -\mathfrak{h} \wedge \sqrt{-1}\rho - \kappa \wedge \mathfrak{h} - [\bar{\theta}] \wedge \bar{\theta},$$
$$d\theta = \theta \wedge \sqrt{-1}\rho - \kappa \wedge \theta - [\bar{\theta}] \wedge \bar{\mathfrak{h}},$$
$$d\kappa = \mathfrak{h} \wedge {}^t\bar{\mathfrak{h}} - \kappa \wedge \kappa + \theta \wedge {}^t\bar{\theta} - [\bar{\theta}] \wedge [\theta].$$

3. Gram-Schmidt orthonormalization process of Spin(7)

To construct the Spin(7)-frame field, we recall the Gram-Schmidt process of G_2-frame.

Lemma 3.1. *For a pair of mutually orthogonal unit vectors e_1, e_4 in Im $\mathfrak{C}$, we put $e_5 = e_1 e_4$. Take a unit vector e_2, which is perpendicular to e_1, e_4 and e_5. If we put $e_3 = e_1 e_2$, $e_6 = e_2 e_4$ and $e_7 = e_3 e_4$ then the matrix*

$$g = [e_1, e_2, e_3, e_4, e_5, e_6, e_7] \in SO(7)$$

is an element of G_2.

3.1. *A method of construction*

By Lemma 3.1, we take a unit vector $e_4 \in \text{Im } \mathfrak{C}$, we can get the G_2-frame field as follows. First, we set

$$N^* = (1/2)(1 - \sqrt{-1}e_4), \quad \bar{N}^* = (1/2)(1 + \sqrt{-1}e_4),$$
$$E_1^* = (1/2)(e_1 - \sqrt{-1}e_5), \quad \bar{E}_1^* = (1/2)(e_1 + \sqrt{-1}e_5),$$
$$E_2^* = (1/2)(e_2 - \sqrt{-1}e_6), \quad \bar{E}_2^* = (1/2)(e_2 + \sqrt{-1}e_6),$$
$$E_3^* = -(1/2)(e_3 - \sqrt{-1}e_7), \quad \bar{E}_3^* = -(1/2)(e_3 + \sqrt{-1}e_7).$$

Next we put

$$(n, f, \bar{n}, \bar{f}) = (N^*, E^*, \bar{N}^*, \bar{E}^*) \begin{pmatrix} P & 0 \\ 0 & \bar{P} \end{pmatrix}$$

for $P \in SU(4) \subset M_{4\times 4}(\mathbf{C})$. Then $(n, f, \bar{n}, \bar{f})$ is the $Spin(7)$-frame.

Remark 3.1. This procedure comes from the following relation

$$Spin(7)/Spin(6) = Spin(7)/SU(4) = S^6 = G_2/SU(3).$$

4. Homogeneous spaces related to $Spin(7)$

In this section, we give a homogeneous spaces which are obtained as the orbit of $Spin(7)$. To do this, we define another type double covering map $\bar{\chi}$ of $Spin(7)$ to $SO(7)$ related to χ, as follows (see [3]).For each $g \in SO(8)$, we set the map $\bar{\chi}(g)(v) \in M_{8\times 8}(R)$ as

$$\bar{\chi}(g)(v) = g \circ R_v \circ g^{-1}$$

where R_v is the right multiplication of octonions defined by $R_v(u) = uv$ for any $u \in \mathfrak{C}$. In general, the set of the image $\bar{\chi}(SO(8))(\mathfrak{C})$ does not coincide with the set $\{R_w \mid w \in \mathfrak{C}\}$. We can give the another definition of $Spin(7)$ by

$$Spin(7) = \{g \in SO(8) \mid \text{there exist } v, w \in \mathfrak{C} \text{ s.t.}, \bar{\chi}(g)(v) = R_w\}.$$

Then we see that

$$w = \chi_g(v).$$

We remark that $\chi_g(1) = 1$ for $1 \in \mathfrak{C}$ and

$$(\chi_g(i), \chi_g(j), \chi_g(k), \chi_g(\varepsilon), \chi_g(i\varepsilon), \chi_g(j\varepsilon), \chi_g(k\varepsilon)), \tag{3}$$

is an $SO(7)$-valued functions on $Spin(7)$. By (1), we have

$$\chi_g(u) = g(u) \times g(1), \tag{4}$$

for any $u \in \text{Im} \, \mathfrak{C}$. First, we recall the following isotropy groups:

$$SU(3) = \{g \in G_2 \mid g(1) = 1, g(\varepsilon) = \varepsilon\}, \tag{5}$$

$$SU(4)(\cong Spin(6)) = \{g \in Spin(7) \mid \chi_g(\varepsilon) = \varepsilon\}, \tag{6}$$

$$Sp(2)(\cong Spin(5)) = \{g \in SU(4) \mid \chi_g(i) = i\}. \tag{7}$$

Then we have the following relations of the inclusions

$$Spin(3)(\cong Sp(1)) \subset Spin(4) \subset Spin(5) \subset Spin(6) \subset Spin(7)$$

and

$$SU(2)(\cong Sp(1)) \subset SU(3) \subset G_2 \subset Spin(7).$$

We also note that $G_2 \cap Spin(6) = G_2 \cap SU(4) = SU(3)$ and $SU(3) \cap Spin(4) = U(1) \times Sp(1)$.

4.1. Spin(7)-orbits

4.1.1. $S^7, S^6, S^5, V_k(\mathfrak{C})$ and $G_k^+(\mathfrak{C})$

Let S^7, S^6 and S^5 be a 7-dimensional unit sphere in $\mathfrak{C}$, a 6-dimensional unit sphere in $\mathrm{Im}\,\mathfrak{C} = \{u \in \mathfrak{C} | \langle u, 1 \rangle = 0\}$, a 5-dimensional unit sphere in $\mathbf{R}^6 = \{u \in \mathrm{Im}\,\mathfrak{C} | \langle u, \varepsilon \rangle = 0\}$, respectively. It is well known that

$$S^7 \cong SO(8)/SO(7) \cong Spin(7)/G_2,$$
$$S^6 \cong SO(7)/SO(6) \cong Spin(7)/Spin(6) \tag{8}$$
$$(= Spin(7)/SU(4)) \cong G_2/SU(3),$$
$$S^5 \cong SO(6)/SO(5) \cong Spin(6)/Spin(5)$$
$$(= SU(4)/Sp(2)) \cong SU(3)/SU(2).$$

Let $V_2(\mathfrak{C})$ be a Stiefel manifold of orthonormal 2-frames in $\mathfrak{C}$. It is well known that

$$V_2(\mathfrak{C}) = \{(u, v) \in S^7 \times S^7 \mid \langle u, v \rangle = 0\}.$$

We shall prove the following

Proposition 4.1.

$$V_2(\mathfrak{C}) \cong Spin(7)/SU(3).$$

Proof. First, we prove that $Spin(7)$ acts transitively on $V_2(\mathfrak{C})$. For any $(u, v) \in V_2(\mathfrak{C})$, by (8), there exists $g \in Spin(7)$ such that $u = g(1)$, where 1 is a multiplicative unit element of $\mathfrak{C}$. Since $\langle u, v \rangle = 0$,

$$v \in T_u^1 S^7 = \{X \in T_u S^7 \mid |X| = 1\}.$$

Here, we identify $T_u S^7$ with $\mathbf{R}^7$, then we have

$$g^{-1}(v) \in T_\varepsilon^1 S^7 \cong S^6.$$

By (8), there exists $h \in G_2 \subset Spin(7)$ such that

$$g^{-1}(v) = h(\varepsilon).$$

Therefore

$$g(h(\varepsilon)) = v. \tag{9}$$

Moreover, since $h(1) = 1$, we get

$$g(h(1)) = g(1) = u. \tag{10}$$

By (9), (10), we have

$$(g(h(1)), g(h(\varepsilon))) = (u, v).$$

Hence $Spin(7)$ acts on $V_2(\mathbb{C})$ transitively, and its isotropy subgroup is $SU(3)$. □

By Proposition 4.1, we can see that

Corollary 4.1.

$$G_2^+(\mathbb{C}) \cong Spin(7)/U(3),$$

where $G_2^+(\mathbb{C})$ be a Grassmann manifold of oriented 2-planes in $\mathbb{C}$. Note that $V_2(\mathbb{C})$ is a principal S^1 fibre bundle over $G_2^+(\mathbb{C})$.

4.1.2. $V_3(\mathbf{R}^8)$ and $G_3^+(\mathbf{R}^8)$

Let $V_3(\mathbf{R}^8)$ and $G_3^+(\mathbf{R}^8)$ be a Stiefel manifold of orthonormal 3-frames in $\mathbf{R}^8$ and a Grassmann manifold of oriented 3-planes in $\mathbf{R}^8$, respectively. In the same way, we can prove the following.

Proposition 4.2.

$$V_3(\mathbf{R}^8) = SO(8)/SO(5) \cong Spin(7)/Sp(1),$$
$$G_3^+(\mathbf{R}^8) = SO(8)/SO(3) \times SO(5) \cong Spin(7)/Sp(1) \times Sp(1)/Z_2.$$

Proof. By the similar argument of the proof in Proposition 4.1, for any $(e_1, e_2, e_3) \in V_3(\mathbf{R}^8)$, there exits $g \in Spin(7)$ such that $g(i) = e_1$, $g(j) = e_2$ and $g(k) = e_3$. Therefore we see that

$$V_3(\mathbf{R}^8) \cong Spin(7)/Sp(1).$$

Also we obtain

$$G_3^+(\mathbf{R}^8) \cong Spin(7)/Sp(1) \times Sp(1)/Z_2.$$

In this case, the action of the isotropy subgroup at $span_{\mathbf{R}}\{i, j, k\} \in G_3^+(\mathbf{R}^8)$ is given by

$$\rho(q_1, q_2)(a + b\varepsilon) = q_1 a \bar{q}_1 + (q_2 b \bar{q}_1)\varepsilon$$

for $(q_1, q_2) \in Sp(1) \times Sp(1)$ and $a + b\varepsilon \in Im\,\mathbb{C}$. Since $\rho(q_1, q_2) = \rho(-q_1, -q_2)$, the isotropy subgroup is $Sp(1) \times Sp(1)/Z_2$. □

Next we consider the canonical form of each element $V \in G_4^+(\boldsymbol{R}^8)$ by $Spin(7)$. We take an orthonormal basis e_1, e_2, e_3, e_4 of V.

(1) If we assume that $e_1(\bar{e}_2 e_3) = e_4$, then there exists $g \in Spin(7)$ satisfying

$$V = span_{\boldsymbol{R}}\{g(1),\ g(i),\ g(j),\ g(k)\}.$$

In this case V is called a Cayley 4-plane.

We note that in [3], the calibrated 4-form Φ on $\mathfrak{C}$ is defined by $\Phi(x, y, z, w) = \langle x, \frac{1}{2}\{y(\bar{z}w) - w(\bar{z}y)\}\rangle$ for any $x, y, z, w \in \mathfrak{C}$. The 4-subspace V of $\mathfrak{C}$ is Cayley 4-plane if and only if $\Phi(V) = 1$.

(2) Suppose that $e_1(\bar{e}_2 e_3) \neq e_4$. We note that there exists $g \in Spin(7)$ such that $g(i) = e_1$, $g(j) = e_2$, $g(k) = e_3$. By the assumption, we may assume that $g(1) \neq e_4$, then we have

$$\dim\big(span_{\boldsymbol{R}}\{g(1), e_4\}\big) = 2.$$

We can take $u \in span_{\boldsymbol{R}}\{g(1), e_4\}$ so that

$$|u| = 1, \langle u, g(1)\rangle = 0.$$

If we put $\langle e_4,\ g(1)\rangle = \cos\theta (0 \leq \theta \leq \pi)$, then

$$e_4 = \cos\theta g(1) + \sin\theta u.$$

Since $u \in (span_{\boldsymbol{R}}\{g(1),\ g(i),\ g(j),\ g(k)\})^{\perp}$, we may put $u = g(\varepsilon)$. Hence we have

$$V = span_{\boldsymbol{R}}\{g(i),\ g(j),\ g(k),\ g(\cos\theta 1 + \sin\theta\varepsilon)\}.$$

Summing up the above arguments, we obtain

Proposition 4.3. *For any $V \in G_4^+(\boldsymbol{R}^8)$, there exist $g \in Spin(7)$ and $\theta \in \boldsymbol{R}$ $(0 \leq \theta \leq \pi)$ satisfying*

$$V = span_{\boldsymbol{R}}\{g(i),\ g(j),\ g(k),\ g(\cos\theta 1 + \sin\theta\varepsilon)\}.$$

4.1.3. $Spin(7)/Sp(1) \times Sp(1) \times Sp(1)/Z_2$ and $G_2^+(\operatorname{Im}\mathfrak{C})$

First, we give a representation of the Lie subgroup of $Spin(7)$ which is related to the Grassmaniann manifolds $G_3^+(\operatorname{Im}\mathfrak{C}) = SO(7)/SO(3) \times SO(4)$ of all oriented 3-dimensional planes in $\boldsymbol{R}^7$. We note that $G_3^+(\operatorname{Im}\mathfrak{C}) \cong G_4^+(\operatorname{Im}\mathfrak{C})$ by using duality. The Lie subgroup $Sp(1) \times Sp(1) \times Sp(1)/Z_2$ corresponds to $SO(3) \times SO(4)$. In fact, we define an action of $Sp(1) \times Sp(1) \times Sp(1)/Z_2$ on $\mathfrak{C}$ as follows

$$\rho(q_1, q_2, q_3)(a + b\varepsilon) = q_2 a\bar{q}_1 + (q_3 b\bar{q}_1)\varepsilon.$$

Then we have

$$
\chi_{\rho(q_1,q_2,q_3)}(a+b\varepsilon) = \rho(q_1,q_2,q_3)\{\rho(\bar{q}_1,\bar{q}_2,\bar{q}_3)(1)(a+b\varepsilon)\}
$$
$$
= \rho(q_1,q_2,q_3)\{\bar{q}_2 q_1(a+b\varepsilon)\} = \rho(q_1,q_2,q_3)\{\bar{q}_2 q_1 a + (b\bar{q}_2 q_1)\varepsilon)\}
$$
$$
= q_1 a\bar{q}_1 + (q_3 b\bar{q}_2)\varepsilon,
$$

for any $a + b\varepsilon \in Im\mathfrak{C}$ and $(q_1, q_2, q_3) \in Sp(1) \times Sp(1) \times Sp(1)$. Therefore $Sp(1) \times Sp(1) \times Sp(1)/Z_2$ is a double covering of $SO(3) \times SO(4)$. In fact, we can easily see that

$$
\rho(q_1, q_2, q_3) = \rho(-q_1, -q_2, -q_3)
$$

and

$$
\chi_{\rho(-q_1,q_2,q_3)} = \chi_{\rho(-q_1,-q_2,-q_3)} = \chi_{\rho(q_1,q_2,q_3)}.
$$

It is known that the homogeneous space $Spin(7)/Sp(1) \times Sp(1) \times Sp(1)/Z_2$ is isomorphic to the Grassmann manifold of all Cayley 4-planes of $\mathfrak{C}$ (in [3]).

Next we give a representation of a Lie subgroup of $Spin(7)$ which is related to the Grassmaniann manifolds $G_2^+(Im\,\mathfrak{C}) = SO(7)/SO(2) \times SO(5)$, of all oriented 2-dimensional planes in $Im\,\mathfrak{C}$. We note that $G_2^+(Im\,\mathfrak{C}) \cong G_5^+(Im\,\mathfrak{C})$ by using duality. We may determine the set

$$
\{g \in Spin(7) \mid \chi_g(i) \wedge \chi_g(\varepsilon) = i \wedge \varepsilon\} = \chi^{-1}(SO(2) \times SO(5)),
$$

which coresponds to the isotropy subgroup at $i \wedge \varepsilon \in G_2^+(Im\,\mathfrak{C})$. From this and (2), we obtain

$$
g = \exp\begin{pmatrix} A & \bar{B} \\ B & \bar{A} \end{pmatrix}
$$

where

$$
A = \begin{pmatrix} \sqrt{-1}\rho_1 & -\bar{\alpha}_1 & -\bar{\alpha}_2 & -\bar{\alpha}_3 \\ \alpha_1 & -\sqrt{-1}\rho_1 & \alpha_3 & \alpha_2 \\ \alpha_2 & -\bar{\alpha}_3 & \sqrt{-1}\rho_2 & -\bar{\beta} \\ \alpha_3 & -\bar{\alpha}_2 & \beta & -\sqrt{-1}\rho_2 \end{pmatrix}
$$

and

$$
B = \begin{pmatrix} 0 & \sqrt{-1}t & 0 & 0 \\ -\sqrt{-1}t & 0 & 0 & 0 \\ 0 & 0 & 0 & -\sqrt{-1}t \\ 0 & 0 & \sqrt{-1}t & 0 \end{pmatrix}
$$

for $\alpha_i, \beta \in \mathbf{C}$ and $\rho_i, t \in \mathbf{R}$. Then $A \in \mathfrak{sp}(2)(\subset \mathfrak{su}(4))$ and $B \in \mathfrak{t}^1$.

4.2. G_2-orbits

4.2.1. $S^6, S^5, V_2(\mathrm{Im}\,\mathfrak{C})$ and $G_2^+(\mathrm{Im}\,\mathfrak{C})$

Let S^6 and S^5 be a 6-dimensional unit sphere in $\mathrm{Im}\,\mathfrak{C}$ and a 5-dimensional unit sphere in $\mathbf{R}^6 = \{u \in \mathrm{Im}\,\mathfrak{C} |\, \langle u, \varepsilon \rangle = 0\}$, respectively. It is well known that

$$S^6 \cong G_2/SU(3), \ \ S^5 \cong SU(3)/SU(2). \tag{11}$$

Let $V_2(\mathrm{Im}\,\mathfrak{C})$ be a Stiefel manifold of oriented 2-frames in $\mathrm{Im}\,\mathfrak{C}$. It is well known that

$$V_2(\mathrm{Im}\,\mathfrak{C}) = \{(u, v) \in S^6 \times S^6 \mid \langle u, v \rangle = 0\}.$$

We shall prove the following

Proposition 4.4.

$$V_2(\mathrm{Im}\,\mathfrak{C}) \cong G_2/SU(2).$$

Proof. First, we prove that G_2 acts transitively on $V_2(\mathrm{Im}\,\mathfrak{C})$. For any $(u, v) \in V_2(\mathrm{Im}\,\mathfrak{C})$, by (11), there exists $g \in G_2$ such that $u = g(\varepsilon)$. Then we get $\langle u, g(i) \rangle = \langle g(\varepsilon), g(i) \rangle = \langle \varepsilon, i \rangle = 0$. Also $\langle u, v \rangle = 0$, we have

$$g(i), \ v \in T_u^1 S^6 = \{X \in T_u S^6 \mid |X| = 1\}.$$

Here, we will identify $T_u S^6$ with $\mathbf{R}^6$, then we obtain

$$i, \ g^{-1}(v) \in T_\varepsilon^1 S^6 \cong S^5.$$

Since $S^5 \cong SU(3)/SU(2)$, there exists $h \in SU(3) \subset G_2$ such that

$$g^{-1}(v) = h(i),$$

where, $SU(3) = \{g \in G_2 \mid g(\varepsilon) = \varepsilon\}$. Therefore

$$g(h(i)) = v. \tag{12}$$

Moreover, since $h(\varepsilon) = \varepsilon$, we get

$$g(h(\varepsilon)) = g(\varepsilon) = u. \tag{13}$$

By (12), (13), we have

$$(g(h(i)), g(h(\varepsilon))) = (u, v).$$

Hence we find G_2 acts on $V_2(\mathrm{Im}\,\mathfrak{C})$ transitively and its isotropy subgroup is $SU(2)$. $\qquad\square$

By Proposition 4.4, we can see that

Corollary 4.2.

$$G_2^+(\mathrm{Im}\,\mathfrak{C}) \cong G_2/U(2),$$

where $G_2^+(\mathrm{Im}\,\mathfrak{C})$ be a Grassmann manifold of oriented 2-planes in $\mathrm{Im}\,\mathfrak{C}$.

4.3. $V_3(R^7)$ and $G_3^+(R^7)$ (G_2-orbit decomposition)

Let $V_3(\mathbf{R}^7)$ and $G_3^+(\mathbf{R}^7)$ be a Stiefel manifold of oriented 3-frames in $\mathbf{R}^7$ and a Grassmann manifold of oriented 3-planes in $\mathbf{R}^7$, respectively. For any $(e_1, e_2, e_3) \in V_3(\mathbf{R}^7)$, by Proposition 4.4, there exits $g \in G_2$ such that $g(i) = e_1$, $g(j) = e_2$. Since $g \in G_2$ we have $g(i)g(j) = g(k)$. In general, $e_1 e_2 \neq e_3$. Therefore we see that two manifolds $V_3(\mathbf{R}^7)$, $G_3^+(\mathbf{R}^7)$ cannot be represented as orbits of G_2.

Next we consider the canonical form of the each element of $G_3^+(\mathbf{R}^7) \ni V$ by G_2. Let $V = span_{\mathbf{R}}\{e_1, e_2, e_3\} \in G_3^+(\mathbf{R}^7)$.

(1) If we assume that $e_1 e_2 = e_3$, then there exists $g \in G_2$ satisfying

$$V = span_{\mathbf{R}}\{g(i),\ g(j),\ g(k)\}.$$

In this case V is called an associative 3-plane.

(2) Suppose that $e_1 e_2 \neq e_3$. We note that there exists $g \in G_2$ such that $g(i) = e_1, g(j) = e_2$. We may assume that $g(k) \neq e_3$. Then we have

$$\dim(span_{\mathbf{R}}\{g(k), e_3\}) = 2.$$

We can take $u \in span_{\mathbf{R}}\{g(k),\ e_3\}$ so that

$$|u| = 1, \langle u, g(k) \rangle = 0.$$

If we put $\langle e_3,\ g(k) \rangle = \cos\theta (0 \leq \theta \leq \pi)$, then

$$e_3 = \cos\theta g(k) + \sin\theta u.$$

Since $u \in (span_{\mathbf{R}}\{g(i),\ g(j),\ g(k)\})^\perp$, we may put $u = g(\varepsilon)$. Hence we have

$$V = span_{\mathbf{R}}\{g(i),\ g(j),\ g(\cos\theta k + \sin\theta\varepsilon)\}.$$

Summing up the above arguments, we obtain

Proposition 4.5. For any $V \in G_3^+(\mathbf{R}^7)$, there exist $g \in G_2$ and $\theta \in \mathbf{R}$ $(0 \leq \theta \leq \pi)$ satisfying

$$V = span_{\mathbf{R}}\{g(i),\ g(j),\ g(\cos\theta k + \sin\theta\varepsilon)\}.$$

A 3-dimensional vector space V in $\mathrm{Im}\mathfrak{C}$ is called *associative* if $\mathrm{span}_{\mathbf{R}}\{u, v, uv\} = V$, where $\{u, v\}$ is an oriented orthonormal pair of V. We also note that the Grassmann manifold $G_{ass}(\mathrm{Im}\mathfrak{C})$ of associative 3-planes are given by

$$G_{ass}(\mathrm{Im}\mathfrak{C}) \simeq G_2/SO(4).$$

We note that the representation

$$\rho_{SO(4)} : SO(4)(\simeq Sp(1) \times Sp(1)/Z_2) \to G_2$$

is given by

$$\rho_{SO(4)}(q_1, q_2)(a + b\varepsilon) = q_1 a \overline{q_1} + (q_2 b \overline{q_1})\varepsilon,$$

where $(q_1, q_2) \in Sp(1) \times Sp(1)$ and $a + b\varepsilon \in \mathrm{Im}\mathfrak{C}$.

4.4. Applications to submanifold theory

In [1] R.L.Bryant showed that any oriented 6-dimensional submanifold $\varphi : M^6 \to \mathfrak{C}$ of the octonions admits an almost complex (Hermitian) structure J defined by

$$\varphi_*(JX) = \varphi_*(X)(\eta \times \xi),$$

where $\{\xi, \eta\}$ is a local oriented orthonormal frame field of the normal bundle of φ over a neighborhood of each point of M^6. The induced almost complex (Hermitian) structure is a $Spin(7)$-invariant in the following sense.

Let $\varphi_1 : M^6 \to \mathfrak{C}$ and $\varphi_2 : N^6 \to \mathfrak{C}$ be two isometric immersions of M^6 and N^6 to the octonions, respectively. If there exists an element $g \in Spin(7)$ and the orientation preserving diffeomorphism ψ such that $g \circ \varphi_1 = \varphi_2 \circ \psi$ (up to a parallel translation), then the two maps are said to be $Spin(7)$-congruent. If the immersions φ_1 and φ_2 are $Spin(7)$-congruent, then the induced orthogonal almost complex structures coincide.

If $g \in SO(8)$, then two maps φ_1, φ_2 are called $SO(8)$-congruent. In this case, the induced orthogonal almost complex structure of $g \circ \varphi_1$ is different from the orginal one of φ_1. We explain this phenomenon. In [6], we give the classification of extrinsic homogeneous almost Hermitian submanifold of the octonions $\mathfrak{C}$ as follows; Let $M^6 = (M^6, \varphi)$ be an extrinsic homogeneous almost Hermitian submanifold of the octonions $\mathfrak{C}$ with respect to the induced almost Hermitian structure induced by the specified isometric immersion φ. Then (M^6, φ) is one of the following 13 submanifolds ((1)–(13)).

(1) φ is totally geodesic. ($\varphi : \mathbf{R}^6 \to \mathfrak{C} \simeq \mathbf{R}^8$, flat Kähler manifold).

(2) φ is totally umbilic in Im$\mathfrak{C}$. ($\varphi : S^6 \to$ Im$\mathfrak{C} \simeq \mathbf{R}^7$, nearly Kähler 6-sphere).

(3) $\varphi : S^1 \times \mathbf{R}^5 \to$ Im$\mathfrak{C} \simeq \mathbf{R}^7$ is defined by

$$\varphi(e^{\sqrt{-1}\theta}, x_1, z_2, z_3)$$
$$= \varepsilon\, x_1 + E_1\, z_1 + E_2\, z_2 + E_3\, e^{\sqrt{-1}\theta} + \overline{E_1\, z_1 + E_2\, z_2 + E_3\, e^{\sqrt{-1}\theta}},$$

for $(e^{i\theta}, x_1, z_1, z_2) \in S^1 \times \mathbf{R} \times \mathbf{C} \times \mathbf{C}\ (\simeq S^1 \times \mathbf{R}^5)$

(4) $\varphi : \mathbf{R}^1 \times \mathbf{R}^5 \to \mathfrak{C}$ is define by ($\gamma : \mathbf{R} \to \mathbf{R}^3$ is a helix)

$$\varphi(t, x_0, x_4, x_5, x_6, x_7)$$
$$= 1\, x_0 + i\ \cos(at) + j\ \sin(at) + k\ bt + \varepsilon\ x_4 + i\varepsilon\ x_5 + j\varepsilon\ x_6 + k\varepsilon\ x_7,$$

for $(t, x_0, x_4, x_5, x_6, x_7) \in \mathbf{R}^1 \times \mathbf{R}^5$.

(5) $\varphi : S^2 \times \mathbf{R}^4 \to$ Im$\mathfrak{C}$ (quasi-Kähler manifold)

$$\varphi(q, x_1, x_2, x_3, x_4) = qi\bar{q} + \varepsilon\ x_1 + i\varepsilon\ x_2 + j\varepsilon\ x_3 + k\varepsilon\ x_4,$$

where $q \in Sp(1) \cong S^3 \subset \mathbf{H}$ and $(x_1, x_2, x_3, x_4) \in \mathbf{R}^4$.

(6) $\varphi : S^3 \times \mathbf{R}^3 \to$ Im$\mathfrak{C}$ is defined by

$$\varphi(q, x_1, x_2, x_3) = i\ x_1 + j\ x_2 + k\ x_3 + q\varepsilon,$$

for $(q, x_1, x_2, x_3) \in S^3 \times \mathbf{R}^3$.

(7) $\varphi : S^5 \times \mathbf{R}^1 \to$ Im$\mathfrak{C}$, is defined by

$$\varphi(x, z_1, z_2, z_3) = \varepsilon\ x + \sum_{i=1}^{3} E_i z_i + \sum_{i=1}^{3} \overline{E_i z_i},$$

where $U \in SU(3)$ and ${}^t(z_1, z_2, z_3) = U\ {}^t(1, 0, 0)$.

(8) $\varphi : S^1 \times S^5 \to S^7 \subset \mathbf{R}^8 \cong \mathfrak{C}$ is defined by

$$\varphi(e^{\sqrt{-1}\theta}, z_1, z_2, z_3) = Ne^{\sqrt{-1}\theta} + e^{(\sqrt{-1}\theta)/3} \sum_{i=1}^{3} E_i z_i$$
$$+ \overline{N}e^{-\sqrt{-1}\theta} + e^{-(\sqrt{-1}\theta)/3} \sum_{i=1}^{3} \overline{E_i z_i},$$

where $U \in SU(3)$ and ${}^t(z_1, z_2, z_3) = U\ {}^t(1, 0, 0)$.

(9) $\varphi : S^3 \times S^3 \to S^7 \subset \mathfrak{C}$ is defined by

$$\varphi(q_1, q_2) = q_1 + q_2\varepsilon,$$

for $(q_1, q_2) \in S^3 \times S^3$.

(10) $\varphi : T^2 \times \mathbf{R}^4 \to \mathfrak{C}$ is defined by

$$\varphi(e^{\sqrt{-1}\theta_1}, e^{\sqrt{-1}\theta_2}, z_1, z_2)$$
$$= N\,e^{\sqrt{-1}\theta_1} + E_1\,e^{\sqrt{-1}\theta_2} + E_2\,z_1 + E_3\,z_2$$
$$+ \overline{N\,e^{\sqrt{-1}\theta_1} + E_1\,e^{\sqrt{-1}\theta_2} + E_2\,z_1 + E_3\,z_2}.$$

(11) $\varphi : \mathbf{R}^2 \times S^1 \times S^3 \to \mathfrak{C}$ is defined by

$$\varphi(x_1, x_2, e^{i\theta}, q) = x_1\,1 + x_2 i + (\cos\theta + i\sin\theta)j(\cos\theta - i\sin\theta)$$
$$+ (q(\cos\theta - i\sin\theta))\varepsilon$$
$$= x_1\,1 + x_2 i + \cos(2\theta)j + \sin(2\theta)k + (q(\cos\theta - i\sin\theta))\varepsilon,$$

where $(x_1, x_2, e^{i\theta}, q) \in \mathbf{R}^2 \times S^1 \times S^3$.

(12) $\tilde{\varphi} : \mathbf{R}^1 \times S^2 \times S^3 \to \mathfrak{C}$ is defined by

$$\varphi(x_1, q_1, q_2) = x_1\,1 + q_1 i \bar{q}_1 + (q_2 \bar{q}_1)\varepsilon,$$

for $(x_1, q_1, q_2) \in \mathbf{R}^1 \times S^3 \times S^3$. The image of $\varphi(\mathbf{R}^1 \times S^3 \times S^3)$ is $\mathbf{R}^1 \times S^2 \times S^3 \subset \mathfrak{C}$.

(13) $\varphi_{t_0} : SO(2) \times SO(3) \times \mathbf{R}^2 \to S^5 \times \mathbf{R}^2 \subset \mathbf{C}^3 \oplus \mathbf{R}^2 = \mathbf{R}^8$ is defined by

$$\varphi_{t_0}(\theta, q, x_1, x_2) = x_1\,1 + x_2\varepsilon$$
$$+ \cos t_0 \Big(\cos\theta q i \bar{q} + \sin\theta(q i \bar{q})\varepsilon \Big)$$
$$+ \sin t_0 \Big(-\sin\theta q j \bar{q} + \cos\theta(q j \bar{q})\varepsilon \Big),$$

for $(\theta, q) \in S^1 \times Sp(1)$, $(x_1, x_2) \in \mathbf{R}^2$ and $0 < t_0 < \pi/4$ (constant).

By Proposition 4.3, we can show in each cases (5), (6), (9), (10), (11) and (12) in above cases, (M^6, φ) admits the 1-parameter family of induced almost complex structures. In other cases, by Proposition 4.1 and Corollary 4.1, the induced almost complex structure is unique under the action of $SO(8)$. To determine the automorphism group of the induced almost Hermitian structures are still open for general 6-dimensional submanifolds in $\mathfrak{C}$. We obtain the automorphism group of the induced almost Hermitian structures of the following hypersurfaces in Im$\mathfrak{C}$.

4.5. Homogeneous hypersurfaces of Im$\mathfrak{C}$

First we note that the induced almost complex structure is unique (up to the action of G_2) in the four homogeneous hypersurfaces of Im$\mathfrak{C}$, $\mathbf{R}^6$, $S^1 \times$

$R^5, R \times S^5, S^6$ with the standard immersion (under the action of $SO(7)$).

(M, φ_k)	$Aut(M, J, g)$	$Iso^+(M)$
R^6	$R^6 \rtimes SU(3)$	$R^6 \rtimes SO(6)$
$S^1 \times R^5$	$U(2) \ltimes R^5$	$SO(2) \times (SO(5) \ltimes R^5)$
$S^5 \times R^1$	$SU(3) \times R^1$	$SO(6) \times R^1$
S^6	G_2	$SO(7)$

4.5.1. *Non-homogeneous induced almost complex structure on* $R^2 \times S^4$

Theorem 4.1. *Let* $\psi_4 : R^2 \times S^4 \hookrightarrow Im\mathfrak{C}$ *be the mapping of* $R^2 \times S^4$ *to* $Im\mathfrak{C}$, *defined by*

$$\psi_4(x_1, x_2, y_0, y_1 q) = y_0 i + x_1 j + x_2 k + y_1 q \varepsilon,$$

where $(x_1, x_2) \in R^2$, $y_0^2 + y_1^2 = 1$, *and* $q \in S^3 \subset H$. *Here we take* S^3 *as a 3-dimensional unit sphere in* H. *Then we have*

$$tr^t \bar{\mathfrak{B}} \mathfrak{B} = \frac{1}{8}(3 + y_0^2), \ tr^t \bar{\mathfrak{A}} \mathfrak{A} = \frac{1}{8} y_1^2,$$

where $\mathfrak{A}, \mathfrak{B}$ *are the* $(2, 0), (1, 1)$-*parts of the second fundamental form. The automorphism group of the induced almost complex structure is* $R^2 \rtimes U(2) (\subset (R^2 \rtimes SO(2)) \times SO(5))$. *Therefore, it does not act transitively on* $R^2 \times S^4$. *The induced almost complex structure on* $R^2 \times S^4$ *(under the action of* $SO(7)$) *is unique up to the action of* G_2.

4.5.2. *1-parameter family of homogeneous almost complex structures on* $S^2 \times R^4$

In this section, we give an explicit representation of G_2-frame fields on $S^2 \times R^4 \subset Im\mathfrak{C}$, and G_2-invariants. Let $q \in S^3 (\subset H)$ be an unit quaternion. We define the map $\pi : S^3 \to S^2$ such that $\pi(q) = q i \bar{q}$, which is called the Hopf map.

Proposition 4.6. *Let* $\varphi_{2,\alpha}$ *be the 1-parameter family of imbeddings of* $S^2 \times R^4$ *to* $Im\mathfrak{C}$. *For each* $\alpha \in [0, \pi/3]$, *we define the imbeddings* $\varphi_{2,\alpha}$ *as*

$$\begin{aligned}
\varphi_{2,\alpha}(q i \bar{q}, \tilde{y}) &= \cos(\alpha) q i \bar{q} + \sin(\alpha)(q i \bar{q})\varepsilon + y_0 \varepsilon \\
&+ y_1(-\sin(\alpha)i + \cos(\alpha)i\varepsilon) \\
&+ y_2(-\sin(\alpha)j + \cos(\alpha)j\varepsilon) \\
&+ y_3(-\sin(\alpha)k + \cos(\alpha)k\varepsilon),
\end{aligned} \tag{14}$$

where $q i \bar{q} \in S^2$ *and* $\tilde{y} = (y_0, y_1, y_2, y_3) \in \mathbf{R}^4$. *Then, we have*

$$tr({}^t\overline{\mathfrak{B}}\mathfrak{B}) = \frac{1}{8}(1 + \cos^2(3\alpha)), \quad tr({}^t\overline{\mathfrak{A}}\mathfrak{A}) = \frac{1}{8}(1 - \cos^2(3\alpha)).$$

The automorphism group of the induced almost Hermitian structure coincides with $SU(2) \ltimes \mathbf{R}^4 (\subset SO(3) \times (SO(4) \ltimes \mathbf{R}^4))$ *and it acts transitively on* $S^2 \times \mathbf{R}^4$ *for any* $\alpha \in (0, \pi/3)$. *In the case,* $\alpha = 0$ *or* $\pi/3$, *the automorphism group of the induced almost Hermitian structure coincides with* $SO(4) \ltimes \mathbf{R}^4 (\subset SO(3) \times (SO(4) \ltimes \mathbf{R}^4))$ *and it acts transitively on* $S^2 \times \mathbf{R}^4$.

From which, we have

Theorem 4.2. *For* $\alpha \in \mathbf{R}$ $(0 \le \alpha \le \pi/3)$, *let* $(S^2 \times \mathbf{R}^4, \varphi_{2,\alpha})$ *be defined as in Proposition 4.6. The family of the imbeddings* $\varphi_{2,\alpha}$ *induce a 1-parameter family of the almost complex structures* J_α *on* $S^2 \times \mathbf{R}^4$, *which are not* G_2−*congruent to each other. Moreover the induced almost Hermitian structure* $(J_\alpha, \langle\,,\,\rangle)$ *is (1,2)-symplectic if and only if* $\alpha = 0$ *or* $\pi/3$.

We here note that $\varphi_{2,\alpha}$ and $\varphi_{2,\alpha+\pi/3}$ are G_2-congruent. The almost Hermitian manifold $(M, J, \langle\,,\,\rangle)$ is said to be *(1,2)-symplectic* if $(d\omega)^{(1,2)} = 0$, where $\omega = \langle J\,,\,\rangle$ is the canonical 2-form (or Kähler form) on M. In our situation, the condition $(d\omega)^{(1,2)} = 0$ is equivalent to the condition $\mathfrak{A} = 0$.

Proposition 4.7. *Let* φ *be any isometric imbedding of* $S^2 \times \mathbf{R}^4$ *to* $Im\mathfrak{C}$. *Then there exist a* $g \in G_2$ *and* $\alpha \in [0, \pi/3]$ *such that* $g \circ \varphi = \varphi_{2,\alpha}$. *Hence the moduli space (up to the action of* G_2*) of isometric imbedddings of* $S^2 \times \mathbf{R}^4$ *to* $Im\mathfrak{C}$ *coincides with* $\{\varphi_{2,\alpha} | \alpha \in [0, \pi/3]\}$.

4.5.3. Deformation of almost complex structures on $S^3 \times \mathbf{R}^3$

We can prove the following

Theorem 4.3. *Let* $\varphi_{3,\alpha} : S^3 \times \mathbf{R}^3 \to Im\mathfrak{C}$ *be a 1-parameter family of imbeddings defined by*

$$\varphi_{3,\alpha}(q_0, q_1, q_2, q_3, x_1, x_2, x_3) = x_1(\cos\alpha i + \sin\alpha\varepsilon) + x_2 j + x_3 k$$
$$+ q_0(-\sin\alpha i + \cos\alpha\varepsilon) + \mathfrak{q}\varepsilon,$$

where

$$\mathfrak{q} = q_1 i + q_2 j + q_3 k, \quad \sum_{i=0}^{3} q_i{}^2 = 1, \quad (x_1, x_2, x_3) \in \mathbf{R}^3$$

and α $(0 \le \alpha \le \pi/2)$ is a parameter of the deformation. Then we have

$$tr(^t\bar{\mathfrak{B}}\mathfrak{B}) = \frac{1}{16}\left(2(1 - q_1{}^2)\sin^2\alpha + 3\right),$$

$$tr(^t\bar{\mathfrak{A}}\mathfrak{A}) = \frac{1}{16}\left(-2(1 - q_1{}^2)\sin^2\alpha + 3\right).$$

Corollary 4.3. *There exists a 1-parameter family of induced almost complex structures J_α $(0 \le \alpha \le \pi/2)$ on $S^3 \times \mathbf{R}^3$, which are not G_2-congruent. Moreover, the induced almost complex structures $J_\alpha(0 < \alpha \le \pi/2)$ are not homogeneous.*

References

1. R. L. Bryant, *Submanifolds and special structures on the octonions*, J. Diff. Geom. 17 (1982), 185–232.
2. T. Fukami and S. Ishihara, *Almost Hermitian structure on S^6*, Tohoku. Math. J. 7 (1955) 151–156.
3. R. Harvey and H.B. Lawson, *Calibrated geometries*, Acta Math. 148 (1982), 47–157.
4. H. Hashimoto, *Characteristic classes of oriented 6-dimensional submanifolds in the octonians*, Kodai Math. J. 16 (1993), 65–73.
5. H. Hashimoto, *Oriented 6-dimensional submanifolds in the octonions III*, Internat. J. Math. and Math. Sci. 18 (1995), 111–120.
6. H. Hashimoto, T. Koda, K. Mashimo and K. Sekigawa, *Extrinsic homogeneous almost Hermitian 6-dimensional submanifolds in the octonions*, Kodai Math. J. 30 (2007), 297–321.
7. H. Hashimoto and K. Mashimo, *On some 3-dimensional CR submanifolds in S^6*, Nagoya Math. J. 156 (1999), 171–185.
8. H. Hashimoto and M. Ohashi, *Orthogonal almost complex structures of hypersurfaces of purely imaginary octonions*, Hokkaido Math. J. 39(2010), 351–387.
9. S. Kobayashi, *Differential geometry of complex vector bundles*, Publications of the mathematical society of Japan 15, Iwanami Shoten, Publishers and Princeton University Press, 1987.
10. S. Kobayashi and K. Nomizu, *Foundations of Differential geometry II*, Wiley-Interscience, New York. 1968.
11. M. Spivak, *A comprehensive introduction to differential geometry IV*, Publish or Perish., 1975.
12. T. Takahashi, *Homogeneous hypersurfaces in space of constant curvature*, J. Math. Soc. Japan 22 (1970), 395–410.
13. R. Takagi and T. Takahashi, *On the principal curvatures of homogeneous hypersurfaces in a sphere*, in *Differential Geometry (in honor of K. Yano)*, Kinokuniya, Tokyo, 1972, 469–481.

Received April 11, 2011
Revised   May 6, 2011

Proceedings of the 2nd International
Colloquium on Differential Geometry
and its Related Fields
Veliko Tarnovo, September 6–10, 2010

177

BÉZIER TYPE ALMOST COMPLEX STRUCTURES ON QUATERNIONIC HERMITIAN VECTOR SPACES

Milen J. HRISTOV*

*Department of Algebra and Geometry, University "St. Cyril and St. Methodius",
5000 Veliko Tarnovo, Bulgaria
E-mail: m.hristov@uni-vt.bg*

The main goal of this article is to give an almost algorithmic way to construct almost complex structures over quaternionic Hermitian vector spaces by means of the construction of a Bézier curve and its related objects.

Keywords: Hopf fibration, Barycentric coordinates, Bézier curves, Almost complex structures, Quaternionic Hermitian vector spaces.

1. Hopf fibration and Bézier curves

1.1. *Hopf fibration and barycentric coordinates*

We recall the following well known definitions:

Definition 1.1. The Hopf map h is defined by

$$h : \mathbb{S}^3 : \sum_{k=1}^{4} x_k^2 = 1 \subset \mathbb{R}^4 \longrightarrow \mathbb{S}^2 : \sum_{s=1}^{3} X_s^2 = 1 \subset \mathbb{R}^3$$
$$\vec{x} = (x_1, x_2, x_3, x_4) \longmapsto \vec{X} = (X_1, X_2, X_3)$$
$$= \left(\vec{x} H_1 \vec{x}^T, \ \vec{x} H_2 \vec{x}^T, \ \vec{x} H_3 \vec{x}^T \right),$$

where the triplet (H_1, H_2, H_3) of 4×4-symmetric matrices is given by

$$H_1 = \begin{pmatrix} 1 & 0 & 0 & 0 \\ 0 & 1 & 0 & 0 \\ 0 & 0 & -1 & 0 \\ 0 & 0 & 0 & -1 \end{pmatrix}, \ H_2 = \begin{pmatrix} 0 & 0 & 0 & 1 \\ 0 & 0 & 1 & 0 \\ 0 & 1 & 0 & 0 \\ 1 & 0 & 0 & 0 \end{pmatrix}, \ H_3 = \begin{pmatrix} 0 & 0 & -1 & 0 \\ 0 & 0 & 0 & 1 \\ -1 & 0 & 0 & 0 \\ 0 & 1 & 0 & 0 \end{pmatrix},$$

and $\vec{x}^T$ denotes the transposed vector of $\vec{x}$.

Definition 1.2. The Hopf fibration is defined by the action of the inverse

---

*Supported by scientific researches fund of "St. Cyril and St. Methodius" University of V. Tarnovo under contract RD-642-01/26.07.2010.

Hopf map $h^{-1} : \mathbb{S}^2 \subset \mathbb{R}^3 \longrightarrow \mathbb{S}^3 \subset \mathbb{R}^4$. The original of each point over $\mathbb{S}^2$ with vector position $\vec{X} = (X_1, X_2, X_3)$ is the circle $h^{-1}(\vec{X})$ over $\mathbb{S}^3$

$$
\text{(I)} \quad h^{-1}\begin{pmatrix} X_1 \\ X_2 \\ X_3 \end{pmatrix} = \begin{pmatrix} x_1 = -\dfrac{1+X_1}{\sqrt{2(1+X_1)}}\sin\theta \\[2mm] x_2 = -\dfrac{1+X_1}{\sqrt{2(1+X_1)}}\cos\theta \\[2mm] x_3 = \dfrac{1}{\sqrt{2(1+X_1)}}(X_2\cos\theta + X_3\sin\theta) \\[2mm] x_4 = \dfrac{1}{\sqrt{2(1+X_1)}}(-X_2\sin\theta + X_3\cos\theta) \end{pmatrix}, \; \theta \in [0, 2\pi].
$$

Thus $\mathbb{S}^3 = \{\{\cos\psi\} \times \sin\psi\, \mathbb{S}^2 : \psi \in [0, \pi]\} \subset \mathbb{R}^4$ is fibered by circles $h^{-1}(\mathbb{S}^2)$. That is, $\mathbb{S}^3$ is a fibered space with base $\mathbb{S}^2$, fiber $\mathbb{S}^1$ and projection h. The quaternionic equation of each such a circle is

$$h^{-1}(\vec{X}): \rho(t) = \frac{1}{\sqrt{2(1+X_1)}}[(1+X_1)\,i + X_2\,j + X_3\,k](\cos\theta + i\sin\theta),\; \theta \in [0, 2\pi].$$

Let $\mathcal{K} = \{O, \vec{a}_0, \vec{a}_1, \vec{a}_2\}$ be an orthonormal coordinate system. We fix the triangle $\triangle A_0A_1A_2$ by $\vec{OA}_k = \vec{a}_k$, $k = 0, 1, 2$ in the plane $\alpha : x + y + z = 1$. Let $\mathbb{S}^2 : x^2 + y^2 + z^2 = 1$ be the unit sphere in $\mathbb{R}^3$ centered at O. We consider the stereographic projection, which is the bijective map

$$\pi : \Sigma_{\mathbb{S}_l^2} \overset{1-1}{\longleftrightarrow} \Sigma_\alpha^*$$

of the set of points $\Sigma_{\mathbb{S}_l^2}$ in lower half-sphere

$$\mathbb{S}_l^2 = \left\{P \in \mathbb{S}^2 : \text{dist}(P, \alpha) \leq \frac{1}{\sqrt{3}} = \text{dist}(O, \alpha)\right\},$$

which is obtained by cutting $\mathbb{S}^2$ with the equatorial plane α' which is parallel to α and contains O (i.e. $\alpha' \parallel \alpha$ and $\alpha' \ni O$), onto the set of points Σ_α^* in the projectively extended euclidean plane $\mathbb{E}_2^*(\alpha)$. Analytically it is expressed as follows:

$$
\text{(II)} \quad
\begin{aligned}
&\pi : \mathbb{S}^2 \ni P(p_1, p_2, p_3) \mapsto P'\left(\frac{p_1}{p}; \frac{p_2}{p}; \frac{p_3}{p}\right) \in \alpha,\; p = \sum_{k=1}^{3} p_k, \\[2mm]
&\pi^{-1} : \alpha \ni Q(q_1, q_2, q_3) \mapsto Q'\left(\frac{q_1}{q}; \frac{q_2}{q}; \frac{q_3}{q}\right) \in \mathbb{S}^2,\; q = |OQ| = \sqrt{\sum_{k=1}^{3} q_k^2}.
\end{aligned}
$$

Thus the barycentric (b-) coordinates of points in α with respect to the triangle $\triangle A_0A_1A_2$ and the cartesian coordinates of the π-corresponding points over $\mathbb{S}^2$ are related. The composition $\pi \circ h$ and its inverse $h^{-1} \circ \pi^{-1}$, which determine the correspondence between $\mathbb{S}^3$ and $\mathbb{E}_2^*(\alpha)$, are well defined and analytically described.

For our later arguments, the problem we stay is the following: For given a Bézier curve with basic $\triangle A_0 A_1 A_2$ and its related geometric objects in $\mathbb{E}_2^*(\alpha)$ [2, p.109], we have to obtain their corresponding images by π^{-1} over $\mathbb{S}^2$ in $\mathbb{R}^3$ and by $h^{-1} \circ \pi^{-1}$ over $\mathbb{S}^3$ in $\mathbb{R}^4$. For this reason we recall needed notions.

The notion of b-trajectories of points is introduced in [2, p.111] in the following way. Given a point P, we denote by (x, y, z) its b-coordinates with respect to the fixed $\triangle A_0 A_1 A_2$. Let λ and μ be the affine ratios which geometrically fix the position of P: If $A_0' = PA_0 \cap A_1 A_2$, then

$$\lambda = (A_2 A_1 A_0') = \frac{\overline{A_2 A_0'}}{\overline{A_1 A_0'}}, \quad \mu = (A_0 A_0' P) = \frac{\overline{A_0 P}}{\overline{A_0' P}}.$$

The b-coordinates (x, y, z) is the unique solution of the system of linear equations

(SLE) $$\begin{cases} z + \lambda y & = 0 \\ (1 - \lambda)y + \mu x & = 0 \\ (1 - \mu)x & = p \end{cases}$$

where $p \neq 0$, $p \in \mathbb{R}$. Its matrix representation is formed as $AX_1 = B$ with

$$A = \begin{pmatrix} 1 & \lambda & 0 \\ 0 & 1 - \lambda & \mu \\ 0 & 0 & 1 - \mu \end{pmatrix}, \quad B = \begin{pmatrix} 0 \\ 0 \\ p \end{pmatrix}, \quad X_1 = \begin{pmatrix} z_1 \\ y_1 \\ x_1 \end{pmatrix}.$$

In order to make the matrix A to be an upper triangular matrix, we place the components of X_1 in reversed order of the barycentric coordinates (x_1, y_1, z_1) of P. The unique solution of (SLE) is

$$X_1 = A^{-1}B = \begin{pmatrix} \dfrac{p\lambda\mu}{(1 - \lambda)(1 - \mu)} \\[2ex] -\dfrac{p\mu}{(1 - \lambda)(1 - \mu)} \\[2ex] \dfrac{p}{1 - \mu} \end{pmatrix}.$$

The triangle matrix A is of stochastic type by columns (or of left-stochastic-like type) and the set $\mathcal{A}$ of all such matrices is a Lie group. By taking the powers A^n, $n \in \mathbb{N}$, the b-trajectory of P which corresponds to X_1^T is the set of points $P_n(x_n, y_n, z_n)$ whose b-coordinates (x_n, y_n, z_n) correspond to the unique solutions of the systems $A^n X_n = B$ with $X_n = \begin{pmatrix} z_n & y_n & x_n \end{pmatrix}^T$. The explicit form for b-coordinates of these points is given in [2, p.111]. For the considered plane α of $\triangle A_0 A_1 A_2$, we shall further assume $p = 1$. We note that the case $p = 0$ determines the unproper (infinity) points, i.e. the unproper (infinity) line ω in $\mathbb{E}_2^*(\alpha)$.

Further we give the notion of the b-exponential map and b-exponential images of points in the following manner. The exponent of $A \in \mathcal{A}$ is

$$
e^A = \begin{pmatrix}
e & e - e^{1-\lambda} & e + \dfrac{\lambda e^{1-\mu} - \mu e^{1-\lambda}}{\mu - \lambda} \\[2ex]
0 & e^{1-\lambda} & \dfrac{\mu \left(-e^{1-\mu} + e^{1-\lambda}\right)}{\mu - \lambda} \\[2ex]
0 & 0 & e^{1-\mu}
\end{pmatrix}.
$$

We set a matrix $\mathbf{e}^A$ by

$$
\mathbf{e}^A := \frac{1}{e} e^A = \begin{pmatrix}
1 & 1 - e^{-\lambda} & 1 + \dfrac{\lambda e^{-\mu} - \mu e^{-\lambda}}{\mu - \lambda} \\[2ex]
0 & e^{-\lambda} & \dfrac{\mu \left(-e^{-\mu} + e^{-\lambda}\right)}{\mu - \lambda} \\[2ex]
0 & 0 & e^{-\mu}
\end{pmatrix} \in \mathcal{A}.
$$

The inverse matrix for $\mathbf{e}^A$ is

$$
(\mathbf{e}^A)^{-1} = e\, e^{-A} = \begin{pmatrix}
1 & 1 - e^{\lambda} & 1 + \dfrac{\mu e^{\lambda} - \lambda e^{\mu}}{\lambda - \mu} \\[2ex]
0 & e^{\lambda} & \dfrac{\mu e^{\mu} - \mu e^{\lambda}}{\lambda - \mu} \\[2ex]
0 & 0 & e^{\mu}
\end{pmatrix} \in \mathcal{A}.
$$

Thus we can define a one-to-one map **b-exp**. For an arbitrary point P whose b-coordinates is given by the solution of (SLE), we correspond it to the point P' whose b-coordinates is given by $(\mathbf{e}^A)^{-1}B$. We call this map a **b-exponential map**. More precisely, we have

$$
P \begin{pmatrix} z \\ y \\ x \end{pmatrix} \equiv \begin{pmatrix} \dfrac{\lambda \mu}{(1-\lambda)(1-\mu)} \\[2ex] -\dfrac{\mu}{(1-\lambda)(1-\mu)} \\[2ex] \dfrac{1}{1-\mu} \end{pmatrix} \xrightarrow{\textbf{b-exp}} P' \begin{pmatrix} z' \\ y' \\ x' \end{pmatrix} \equiv \begin{pmatrix} 1 + \dfrac{\mu e^{\lambda} - \lambda e^{\mu}}{\lambda - \mu} \\[2ex] \dfrac{\mu e^{\mu} - \mu e^{\lambda}}{\lambda - \mu} \\[2ex] e^{\mu} \end{pmatrix},
$$

and the affine ratios λ' and μ' which fix the position of P' are given as

$$\lambda' = -\frac{z'}{y'} = \frac{\lambda - \mu + \mu\,e^\lambda - \lambda\,e^\mu}{\mu(e^\lambda - e^\mu)} = (A_2 A_1 A_e),$$

$$\mu' = -\frac{y' + z'}{x'} = 1 - e^{-\mu} = (A_0 A_e P'),$$

where $A_e = A_0 P' \cap A_1 A_2$. With the action of π^{-1} in (II) we can obtain $(\mathbb{S}^2\text{-})$sphere images corresponding to points P, P_n, $P' = \mathbf{b\text{-}exp}(P)$. Similarly, with $h^{-1} \circ \pi^{-1}$ in (I) we can obtain Hopf fibers (circles) over $\mathbb{S}^3$ corresponding to these points.

1.2. *The analogues in a real projective space*

Let $\mathbb{E}_3^*$ be the projectively extended euclidean space $\mathbb{R}^3$. It is embedded in $\mathbb{R}^4$ as the real hyperplane $x + y + z + u = 1$ with respect to an othonormal coordinate system $\mathcal{K} = \{O, \vec{a}_0, \vec{a}_1, \vec{a}_2, \vec{a}_3\}$. Let $A_0 A_1 A_2 A_3$ be the tetrahedron in $\mathbb{R}^3$ generated by $\overrightarrow{OA_k} = \vec{a}_k$, $k = 0, 1, 2, 3$. Thus for an arbitrary point $P \in \mathbb{E}_3^*$ there exist points $P_0 := A_0 P \cap p.(A_1 A_2 A_3)$, $P_1' := A_1 P_0 \cap A_2 A_3$ and the affine ratios $\lambda = (A_2 A_3 P_1')$, $\mu = (A_1 P_1' P_0)$, $\nu = (A_0 P_0 P)$ which fix the position of P. Moreover, for an arbitrary $P \in \mathbb{E}_3^*$ there exists unique $(x, y, z, u) \in \mathbb{R}^4$ satisfying

$$\begin{cases} x + y + z + u = 1, \\ \overrightarrow{OP} = x\,\overrightarrow{OA_0} + y\,\overrightarrow{OA_1} + z\,\overrightarrow{OA_2} + u\,\overrightarrow{OA_3}. \end{cases}$$

It is called the barycentric coordinates of $P(x, y, z, u)$. This quadruplet (x, y, z, u) is the unique solution of the following system of linear equations

$$\begin{cases} u + \lambda z + \qquad + \qquad = 0, \\ (1 - \lambda)z + \mu y + \qquad = 0, \\ (1 - \mu)y + \nu x = 0, \\ (1 - \nu)x = 1. \end{cases}$$

Its matrix representation is formed as $AX_1 = B$ with

$$A = \begin{pmatrix} 1 & \lambda & 0 & 0 \\ 0 & 1 - \lambda & \mu & 0 \\ 0 & 0 & 1 - \mu & \nu \\ 0 & 0 & 0 & 1 - \nu \end{pmatrix}, \quad B = \begin{pmatrix} 0 \\ 0 \\ 0 \\ 1 \end{pmatrix}, \quad X_1 = \begin{pmatrix} u_1 \\ z_1 \\ y_1 \\ x_1 \end{pmatrix}.$$

In order to make the matrix A to be an upper triangular matrix, we place the components of X_1 in reversed order of the barycentric coordinates (x_1, y_1, z_1, u_1) of P. The set $\mathcal{A}$ of all such left-stochastic-like matrices is a multiplicative non-commutative Lie group. By calculating the inverse matrix we have

$$
A^{-1} = \begin{pmatrix}
1 & \dfrac{-\lambda}{1-\lambda} & \dfrac{\lambda\mu}{(1-\lambda)(1-\mu)} & \dfrac{-\lambda\mu\nu}{(1-\lambda)(1-\mu)(1-\nu)} \\[2mm]
0 & \dfrac{1}{1-\lambda} & \dfrac{-\mu}{(1-\lambda)(1-\mu)} & \dfrac{\mu\nu}{(1-\lambda)(1-\mu)(1-\nu)} \\[2mm]
0 & 0 & \dfrac{1}{1-\mu} & \dfrac{-\nu}{(1-\mu)(1-\nu)} \\[2mm]
0 & 0 & 0 & \dfrac{1}{1-\nu}
\end{pmatrix},
$$

hence we find the b-coordinates of the point P which corresponds to X_1^T is of the form

$$
P\left(\frac{-\lambda\mu\nu}{(1-\lambda)(1-\mu)(1-\nu)};\ \frac{\mu\nu}{(1-\lambda)(1-\mu)(1-\nu)};\ \frac{-\nu}{(1-\mu)(1-\nu)};\ \frac{1}{1-\nu}\right).
$$

We take vectors $X_n = (u_n, z_n, y_n x_n)^T$ with $A^n X_n = B$ for $n = 1, 2, \ldots$. The set of points P_n, $n = 1, 2, \ldots$, whose b-coordinates are (x_n, y_n, z_n, u_u) is said to be the *barycentric trajectory* of this point P. After strong calculations we obtain that the b-coordinates of this b-trajectory are of the form

$$
X_n = \begin{pmatrix} u_n \\ z_n \\ y_n \\ x_n \end{pmatrix} = \begin{pmatrix}
1 - \displaystyle\mathop{\sigma}_{(\lambda,\mu,\nu)}\left\{\dfrac{\mu\nu}{(1-\lambda)^n(\lambda-\mu)(\lambda-\nu)}\right\} \\[4mm]
\mu\nu\left[\displaystyle\mathop{\sigma}_{(\lambda,\mu,\nu)}\left\{\dfrac{1}{(1-\lambda)^n(\lambda-\mu)(\lambda-\nu)}\right\}\right] \\[4mm]
\dfrac{\nu}{\nu-\mu}\left[\dfrac{1}{(1-\mu)^n} - \dfrac{1}{(1-\nu)^n}\right] \\[4mm]
\dfrac{1}{(1-\nu)^n}
\end{pmatrix},
$$

$n = 1, 2, \ldots$, where $\displaystyle\mathop{\sigma}_{(\lambda,\mu,\nu)}\{\ldots\}$ denotes the cyclic summation on (λ, μ, ν).

Up to a constant multiplier the exponent of the matrix A is also of left-stochastic-like type:

$$\mathbf{e}^A = \frac{1}{e}\,e^A = \begin{pmatrix} 1 & 1-e^{-\lambda} & 1-\dfrac{\lambda e^{-\mu}-\mu e^{-\lambda}}{\lambda-\mu} & 1-\underset{(\lambda,\mu,\nu)}{\sigma}\left\{\dfrac{\mu\nu e^{-\lambda}}{(\lambda-\mu)(\lambda-\nu)}\right\} \\[3ex] 0 & e^{-\lambda} & \dfrac{\mu(e^{-\mu}-e^{-\lambda})}{\lambda-\mu} & \dfrac{\mu\nu\,\underset{(\lambda,\mu,\nu)}{\sigma}\{\lambda(e^{-\nu}-e^{-\mu})\}}{(\lambda-\mu)(\mu-\nu)(\lambda-\nu)} \\[3ex] 0 & 0 & e^{-\mu} & \dfrac{\nu(e^{-\nu}-e^{-\mu})}{\mu-\nu} \\[3ex] 0 & 0 & 0 & e^{-\nu} \end{pmatrix}.$$

Analogously to the projective-plane-case there exists the b-exponential map in the projective-space-case. The action of h in (I) and $\pi \circ h$ in (II) over each point in the section $\mathbb{S}^3 \cap \mathbb{E}_3^* : x + y + z + u = 1$ give images over $\mathbb{S}^2$ and $\mathbb{E}_2^*$, respectively.

1.3. Bézier curves and their related geometric objects

Definition 1.3. A (standard) Bézier curve of nth power with basic simplex $A_0 A_1 A_2 \ldots A_n$ in the euclidean space $\mathbb{R}^{n+1}$, endowed with the standard orthonormal coordinate system $\{O, \{\vec{e}_k\}_{k=0}^n\}$, is the linear combination of the vectors $\vec{a}_k = \overrightarrow{OA_k}$ whose coefficients are the basic Bernstein polynomials of nth power $B_{(n,k)} = \binom{n}{k} t^k (1-t)^{n-k}$, $k = 0, 1, 2, \ldots n$, for $t \in [0,1]$.

So, such a curve is described as the vector equation

$$\mathcal{C}_{\{A_0 A_1 \ldots A_n\}} : \vec{r}(t) = \sum_{k=0}^n \binom{n}{k} t^k (1-t)^{n-k} \vec{a}_k, \qquad t \in [0,1].$$

Moreover, if we consider the vector-valued function $\overrightarrow{B}_n(t) = \{B_{(n,0)}(t), B_{(n,1)}(t), \ldots B_{(n,n)}(t)\}$ on $[0,1]$ and fix a non-zero constant vector $\vec{w} = (w_0, w_1, \ldots, w_n) \in \mathbb{R}^{n+1}$ which is not orthogonal to $\overrightarrow{B}_n(t)$, then we get a curve

$$\mathcal{C}^{\vec{w}}_{\{A_0 A_1 \ldots A_n\}} : \vec{r}(t) = \frac{1}{B_n^w(t)} \sum_{k=0}^n w_k \cdot \binom{n}{k} t^k (1-t)^{n-k} . \vec{a}_k, \qquad t \in [0,1],$$

where $B_n^w(t) = \vec{w} \cdot \overrightarrow{B}_n(t)$ with the euclidean scalar product. This is called a **rational Bézier curve** of nth power with basic simplex $A_0 A_1 \ldots A_n$

and with weight-vector $\vec{w}$. When the vectors $\vec{w}$ and $\overrightarrow{B}_n(t)$ are linearly independent on $[0,1]$, then by means of the 2-fold vector cross product $\overrightarrow{\mathcal{B}}_n(t) = \overrightarrow{B}_n(t) \times \vec{w}$ in $\mathbb{R}^{n+1}$ for the cases $n = 2$ or $n = 6$ [1, p.227, Theorem (4.1)], there exist **cross product rational Bézier curves** in $\mathbb{R}^3$ and in $\mathbb{R}^7$.

Further we shall deal basically with the case $n = 2$. Let $\triangle A_0 A_1 A_2$ be a fixed triangle in the plane $\alpha : x + y + z = 1$ which is projectively extended in $\mathbb{R}^3$ as in the subsection 1.1. In the sense of the b-coordinates, the standard Bézier curve $\mathcal{C}_{\{A_0 A_1 A_2\}}$ and the rational one $\mathcal{C}_{\{A_0 A_1 A_2\}}^{\vec{w}}$ with weight-vector $\vec{w} = (w_0, w_1, w_2) \in \mathbb{R}^3$ can be considered as the following locuces respectively:

$$\mathcal{C}_{\{A_0 A_1 A_2\}} = \left\{ P(B_{(2,0)}(t), B_{(2,1)}(t), B_{(2,2)}(t)), t \in [0,1] \right\},$$

$$\mathcal{C}_{\{A_0 A_1 A_2\}}^{\vec{w}} = \left\{ P\left(\frac{w_0 . B_{(2,0)}(t)}{B_2^w(t)}, \frac{w_1 . B_{(2,1)}(t)}{B_2^w(t)}, \frac{w_2 . B_{(2,2)}(t)}{B_2^w(t)} \right), t \in [0,1] \right\}.$$

Theorem 1.1. *For a rational Bézier curve* $\mathcal{C}_{\{A_0 A_1 A_2\}}^{\vec{w}}$, *the following matrix representations hold related with the Bernstein polynomial bases and with power polynomial bases:*

(i) $\mathcal{C}_{\{A_0 A_1 A_2\}}^{\vec{w}} : \vec{r}(t) = \dfrac{1}{B_2^w(t)} \left(w_0 B_{(2,0)}(t), \ w_1 B_{(2,1)}(t), \ w_2 B_{(2,2)}(t) \right) \begin{pmatrix} \vec{a}_0 \\ \vec{a}_1 \\ \vec{a}_2 \end{pmatrix} ;$

(ii) $\mathcal{C}_{\{A_0 A_1 A_2\}}^{\vec{w}} : \vec{r}(t) = \dfrac{1}{B_2^w(t)} \left(1, \ t, \ t^2 \right) \cdot T \cdot \begin{pmatrix} \vec{a}_0 \\ \vec{a}_1 \\ \vec{a}_2 \end{pmatrix} ,$

with the transport matrix

$$T = XM = \begin{pmatrix} w_0 & 0 & 0 \\ -2w_0 & 2w_1 & 0 \\ w_0 & -2w_1 & w_2 \end{pmatrix},$$

where

$$X = \begin{pmatrix} w_0 & 0 & 0 \\ -2(w_0 - w_1) & w_1 & 0 \\ w_0 - 2w_1 + w_2 & w_2 - w_1 & w_2 \end{pmatrix}, \quad M = \begin{pmatrix} 1 & 0 & 0 \\ -2 & 2 & 0 \\ 1 & -2 & 1 \end{pmatrix} ;$$

(iii) $\mathcal{C}_{\{A_0 A_1 A_2\}}^{\vec{w}} : \vec{r}(t) = \dfrac{1}{B_2^w(t)} \left(1, \ t, \ t^2 \right) \cdot X \cdot \begin{pmatrix} \overrightarrow{OA_0} \\ 2\overrightarrow{A_0 A_1} \\ \overrightarrow{A_1 A_0} + \overrightarrow{A_1 A_2} \end{pmatrix} .$

The scalar product $B_2^w(t) = \vec{w} \cdot \vec{B}(t)$ is expressed as

$$B_2^w(t) = w_0(1-t)^2 + w_1 2t(1-t) + w_2 t^2$$
$$= (w_0 - 2w_1 + w_2)t^2 - 2(w_0 - w_1)t + w_0.$$

When the vectors $\vec{w}$ and $\vec{B}_2(t)$ are linearly independent on $[0,1]$, then the vector cross product

$$\vec{\mathcal{B}}_2(t) = \vec{B}_2(t) \times \vec{w} = \Big(\mathcal{B}_{(2,0)}(t), \mathcal{B}_{(2,1)}(t), \mathcal{B}_{(2,2)}(t)\Big)$$

with

$$\mathcal{B}_{(2,0)}(t) = \begin{vmatrix} B_{(2,1)} & B_{(2,2)} \\ w_1 & w_2 \end{vmatrix}, \quad \mathcal{B}_{(2,1)}(t) = \begin{vmatrix} B_{(2,2)} & B_{(2,0)} \\ w_2 & w_0 \end{vmatrix}, \quad \mathcal{B}_{(2,2)}(t) = \begin{vmatrix} B_{(2,0)} & B_{(2,1)} \\ w_0 & w_1 \end{vmatrix}$$

defines on $[0,1]$ the **cross product rational Bézier curve** for $\triangle A_0 A_1 A_2$ by

$$C^{*\,\vec{w}}_{\{A_0 A_1 A_2\}} : \vec{r}(t) = \frac{1}{\mathcal{B}_2(t)} \sum_{k=0}^2 \mathcal{B}_{(2,k)}(t)\vec{a}_k, \quad \mathcal{B}_2(t) = \sum_{k=0}^2 \mathcal{B}_{(2,k)}(t) \neq 0.$$

From the view point of b-coordinates description, we have

$$C^{*\,\vec{w}}_{\{A_0 A_1 A_2\}} = \left\{ P\Big(\frac{\mathcal{B}_{(2,0)}(t)}{\mathcal{B}_2(t)}, \frac{\mathcal{B}_{(2,1)}(t)}{\mathcal{B}_2(t)}, \frac{\mathcal{B}_{(2,2)}(t)}{\mathcal{B}_2(t)}\Big), \ t \in [0,1] \right\}.$$

Theorem 1.2. *The matrix form of the equation of $C^{*\,\vec{w}}_{\{A_0 A_1 A_2\}}$ is*

$$C^{*\,\vec{w}}_{\{A_0 A_1 A_2\}} : \vec{r}(t) = \frac{1}{\mathcal{B}_2(t)} \big(\mathcal{B}_{(2,0)}, \mathcal{B}_{(2,1)}, \mathcal{B}_{(2,2)}\big) \begin{pmatrix} \vec{a}_0 \\ \vec{a}_1 \\ \vec{a}_2 \end{pmatrix}.$$

(i) *Its representation with the Bernstein basis is*

$$C^{*\,\vec{w}}_{\{A_0 A_1 A_2\}} : \vec{r}(t) = \frac{1}{\mathcal{B}_2(t)} \big(B_{(2,0)}, B_{(2,1)}, B_{(2,2)}\big) \begin{pmatrix} 0 & -w_2 & w_1 \\ w_2 & 0 & -w_0 \\ -w_1 & w_0 & 0 \end{pmatrix} \begin{pmatrix} \vec{a}_0 \\ \vec{a}_1 \\ \vec{a}_2 \end{pmatrix}.$$

(ii) *Its representation with power basis is*

$$C^{*\,\vec{w}}_{\{A_0 A_1 A_2\}} : \vec{r}(t) = \frac{1}{\mathcal{B}_2(t)} \big(1, t, t^2\big) \begin{pmatrix} 0 & -w_2 & w_1 \\ 2w_2 & 2w_2 & -2(w_0 + w_1) \\ -w_1 - 2w_2 & w_0 - w_2 & w_1 + 2w_0 \end{pmatrix} \begin{pmatrix} \vec{a}_0 \\ \vec{a}_1 \\ \vec{a}_2 \end{pmatrix}.$$

Remark 1.1. The representations in Theorem 1.1 and in Theorem 1.2 make it easy to calculate the derivatives $\frac{d\vec{r}}{dt} = \dot{\vec{r}}$, $\frac{d^2\vec{r}}{dt^2} = \ddot{\vec{r}}$, ..., and hence make us easy to calculate the invariants and to obtain a geometric information for the corresponding curve.

Remark 1.2. The choice $w_0 = w_1 = w_2 = 1$ makes rational Bézier curves to be standard Bézier curves. In this case the curve $\mathcal{C}^* {}^{\vec{w}}_{\{A_0 A_1 A_2\}}$ does not exists. It coincides with the infinity line, because of $\mathcal{B}_2(t) \equiv 0$.

Remark 1.3. When we choose the weight-vector as $\vec{w} = (w_0 = 1,\ w_1 = u,\ w_2 = 1)$, the number $u \in \mathbb{R}$ is called normal parameterization for $\mathcal{C}^{\vec{w}}_{\{A_0 A_1 A_2\}}$. In this case we shall denote $\mathcal{C}^{\vec{w}}_{\{A_0 A_1 A_2\}}$ as $\overset{(u)}{\mathcal{C}}{}_{\{A_0 A_1 A_2\}}$, $B_2^w(t)$ as $B_2(u,t)$, $\mathcal{C}^* {}^{\vec{w}}_{\{A_0 A_1 A_2\}}$ as $\overset{(u)}{\mathcal{C}^*}{}_{\{A_0 A_1 A_2\}}$ and $\mathcal{B}_2(t)$ as $\mathcal{B}_2(u,t)$. Thus we have

$$(1) \quad \begin{cases} \overset{(u)}{\mathcal{C}}{}_{\{A_0 A_1 A_2\}} : \vec{r}(t) = \frac{1}{B_2(u,t)}\Big((1-t)^2,\ 2ut(1-t),\ t^2\Big)\begin{pmatrix}\vec{a}_0\\ \vec{a}_1\\ \vec{a}_2\end{pmatrix}, \\[2mm] B_2(u,t) = 2(1-u)t^2 - 2(1-u)t + 1, \quad t \in [0,1],\ u \in \mathbb{R}, \end{cases}$$

$$\begin{cases} \overset{(u)}{\mathcal{C}^*}{}_{\{A_0 A_1 A_2\}} : \vec{r}(t) = \frac{1}{\mathcal{B}_2(u,t)}\Big(2t - (u+2)t^2,\ 2t - 1,\ 2t(t-1) + u(1-t)^2\Big)\begin{pmatrix}\vec{a}_0\\ \vec{a}_1\\ \vec{a}_2\end{pmatrix}, \\[2mm] \mathcal{B}_2(u,t) = (2t-1)(1-u), \quad t \in [0,1],\ u \in \mathbb{R}\backslash\{1\}. \end{cases}$$

The geometric meaning of the normal parameterization is well known. The type of the curve depends on the number of the infinity points which are concurrent with the same one. That is, it depends on the number of solutions (t_k) for the equation $B_2(u,t) = 0$ and $\mathcal{B}_2(u,t) = 0$, respectively. Concretely, for the curve $\overset{(u)}{\mathcal{C}}{}_{\{A_0 A_1 A_2\}}$ in (1), the discriminant of $B_2(u,t)$ is $D_{B_2} = 4(u^2 - 1)$. Hence we find that the type of this curve is

  i) elliptic, when $u \in (-1, 1)$,

  ii) hyperbolic, when $u \in (-\infty, -1) \cup (1, +\infty)$,

  iii) parabolic, when $u = \pm 1$.

The curve passes through the end-points $A_0(t = 0)$ and $A_2(t = 1)$. By the equation (1), we find that u shows the position of the arc $\widehat{A_0 A_2}$ of $\overset{(u)}{\mathcal{C}}{}_{\{A_0 A_1 A_2\}}$. It is contained in the interior $\text{int}\triangle A_0 A_1 A_2$ of the basic triangle if and only if $u \in (0, +\infty)$. When $u \in (-\infty, 0)$, it is contained in the exterior $\text{ext}\triangle A_0 A_1 A_2$ of $\triangle A_0 A_1 A_2$. When $u = 0$, it is the segment $A_0 A_2$. For any point C on the segment $A_0 A_2$, the line $A_1 C$ meets with the curve $\overset{(u)}{\mathcal{C}}{}_{\{A_0 A_1 A_2\}}$ at two points. We denote them by $B \in \text{int}\triangle A_0 A_1 A_2$ and $D \in \text{ext}\triangle A_0 A_1 A_2$. Then the quadruplet $\{A_1, C, B, D\}$ is a harmonic quadruplet, that is, their affine ratios satisfy $(A_1 CB) = -(A_1 CD)$. In particular, when C coincides with the midpoint M of the segment $A_0 A_2$, then $B \equiv P(t = \frac{1}{2})$ and $D \equiv Q(t = -\frac{1}{2})$. In this case the affine ratios $(A_1 MP) = -(A_1 MQ) = -1/u$

give the geometric meaning of the parameter u. Moreover, we get the first derivative of this curve as follows:

$$\dot{\vec{r}}(t) = \frac{2}{(B_2(u,t))^2}\left((1-t)\begin{vmatrix} u & 1 \\ t & t-1 \end{vmatrix}, u(1-2t), t\begin{vmatrix} u & 1 \\ t-1 & t \end{vmatrix}\right)\begin{pmatrix} \vec{a}_0 \\ \vec{a}_1 \\ \vec{a}_2 \end{pmatrix}.$$

The tangent vectors at the end points are

$$\vec{t}_{A_0} = \dot{\vec{r}}(0) = 2u\overrightarrow{A_0A_1} \qquad \text{and} \qquad \vec{t}_{A_2} = \dot{\vec{r}}(1) = 2u\overrightarrow{A_1A_2}.$$

The tangent vector of $\overset{(u)}{\mathcal{C}}_{\{A_0A_1A_2\}}$ at the point $P(t=\frac{1}{2})$ is parallel to A_0A_2. At the points $P(t=\frac{1}{2})$ and $Q(t=-\frac{1}{2})$ we have the following:

$$\vec{t}_P = \dot{\vec{r}}(\frac{1}{2}) = \frac{2}{u+1}\overrightarrow{A_0A_2},$$

$$\vec{t}_Q = \dot{\vec{r}}(-\frac{1}{2}) = \frac{2}{(3u-5)^2}\{(u-3)\overrightarrow{A_0A_2} + 8u\overrightarrow{A_0A_1}\}.$$

By this information we get the well known conclusion: The shape of the curve $\overset{(u)}{\mathcal{C}}_{\{A_0A_1A_2\}}$ is

  i) ellipse, when $u \in (-1, 1)$,
 ii) hyperbola, when $u \in (1, +\infty)$,
iii) parabola, when $u = +1$,
iv) degenerating, when $u \in (-\infty, -1]$.

We shall discuss about cross product curve $\overset{(u)}{\mathcal{C}^*}_{\{A_0A_1A_2\}}$ in different articles.

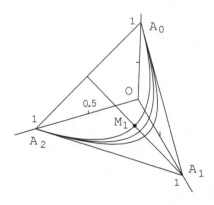

Fig. 1. B-trajectories of a Bézier parabola

We take a rational Bézier curve $\overset{(u)}{\mathcal{C}}_{\{A_0A_1A_2\}}$ with normal parameterization. Obviously $\overset{(1)}{\mathcal{C}}_{\{A_0A_1A_2\}} \equiv \mathcal{C}_{\{A_0A_1A_2\}}$ is a standard Bézier parabola. Its b-trajectories $\mathcal{C}^{(n)}_{\{A_0A_1A_2\}}$ (Fig.1) and their corresponding envelopes $\mathcal{E}_n$ for the sets of polar lines with respect to $\triangle A_0A_1A_2$ are discussed in [2, pp.116-118]. We can apply these ideas to $\overset{(u)}{\mathcal{C}}_{\{A_0A_1A_2\}}$. In addition, we obtain the b-exponential image for $\mathcal{C}_{\{A_0A_1A_2\}}$ (Fig.2) by means of the locus of the b-exponential images of each its point.

Theorem 1.3. *The b-exponential image of the standard Bézier parabola is*

$$\mathcal{C}^{exp}_{\{A_0A_1A_2\}} : \vec{r}(t) = \left(e^\mu , \frac{\mu(e^\mu - e^\lambda)}{\lambda - \mu} , 1 + \frac{\mu\, e^\lambda - \lambda\, e^\mu}{\lambda - \mu} \right) \begin{pmatrix} \vec{a}_0 \\ \vec{a}_1 \\ \vec{a}_2 \end{pmatrix} , \quad t \in [0,1],$$

where

$$\lambda = \frac{t}{2(t-1)} , \quad \mu = \frac{t(t-2)}{(t-1)^2} , \quad \vec{r}(0) = \vec{a}_0 , \quad \vec{r}(1) = \lim_{\substack{t\to 1 \\ t<1}} \vec{r}(t) = \vec{a}_2.$$

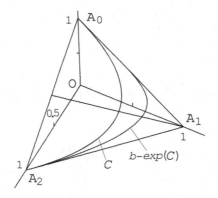

Fig. 2. B-exponential image of the Bézier parabola

1.4. *The sphere image and the Hopf fibration of a rational Bézier curve with normal parameterization*

The following statement is well known:

Proposition 1.1. *For arbitrary fixed two triangles $\triangle$ and $\triangle'$ in an extended euclidean plane, there exists a unique affine transformation φ which*

maps $\triangle$ *onto* $\triangle'$ $\bigl(i.e.\ \triangle' = \varphi(\triangle)\bigr)$. *Moreover* φ *maps a Bézier curve* $\mathcal{C}_\triangle$
for $\triangle$ *onto the Bézier curve for* $\triangle'$: $\varphi(\mathcal{C}_\triangle) = \mathcal{C}_{\varphi(\triangle)}$.

Thus, we see $\overset{(u)}{\mathcal{C}}_{\{A_0 A_1 A_2\}}$ is affine invariant. Therefore, we only consider
the case that the basic triangle $\triangle A_0 A_1 A_2$ in $\alpha : x + y + z = 1$ is equilateral,
and the ordered triplet $\{\vec{a}_0 = \overrightarrow{OA_0}, \vec{a}_1 = \overrightarrow{OA_1}, \vec{a}_2 = \overrightarrow{OA_2}\}$, where O is
the center of the unit sphere $\mathbb{S}^2$, is orthonormal. The image of the basic
$\triangle A_0 A_1 A_2$ over the lower half-sphere $\mathbb{S}^2_l$ of $\mathbb{S}^2$ is the geodesic triangle, formed
by the arcs $\overset{\frown}{A_i A_{i+1}}$. We denote it like $\triangle A_0 A_1 A_2{}_{|\mathbb{S}^2_l}$. We denote the image of
the curve $\overset{(u)}{\mathcal{C}}_{\{A_0 A_1 A_2\}}$ over $\mathbb{S}^2_l$, i.e. $\pi^{-1}(\overset{(u)}{\mathcal{C}}_{\{A_0 A_1 A_2\}})$ like $\overset{(u)}{\mathcal{C}}_{|\mathbb{S}^2}$ and the Hopf
fibees of $\overset{(u)}{\mathcal{C}}_{|\mathbb{S}^2}$, i.e. $h^{-1}(\overset{(u)}{\mathcal{C}}_{|\mathbb{S}^2})$ like $\overset{(u)}{\mathcal{C}}_{|\mathbb{S}^3}$. The following Theorem follows
immediately from (1), (II) and the Hopf fibration (I).

Theorem 1.4. *Let* $\overset{(u)}{\mathcal{C}}_{\{A_0 A_1 A_2\}}$ *be a normally parameterized rational Bézier
curve in the interior of the basic* $\triangle A_0 A_1 A_2$. *Then we have the following:*

(1) *Its sphere* ($\mathbb{S}^2$-*)image is the curve*

$$\overset{(u)}{\mathcal{C}}_{|\mathbb{S}^2} : \vec{r}(t) = e^{-\sigma(u,t)}\left((1-t)^2,\ 2ut(1-t),\ t^2\right)\begin{pmatrix}\vec{a}_0\\ \vec{a}_1\\ \vec{a}_2\end{pmatrix};$$

(2) *Its set of Hopf fibers in* $\mathbb{S}^3$ *is*

$$\overset{(u)}{\mathcal{C}}_{|\mathbb{S}^3} : (x_0,\ x_1,\ x_2,\ x_3)^T$$
$$= \left\{\frac{-e^{\frac{\sigma}{2}}}{\sqrt{2(e^\sigma + (1-t)^2)}}\left[e^{-\sigma}\begin{pmatrix}\sin\theta & 0 & 0\\ \cos\theta & 0 & 0\\ 0 & -\cos\theta & -\sin\theta\\ 0 & \sin\theta & -\cos\theta\end{pmatrix}\begin{pmatrix}(1-t)^2\\ 2ut(1-t)\\ t^2\end{pmatrix} + \begin{pmatrix}\sin\theta\\ \cos\theta\\ 0\\ 0\end{pmatrix}\right] : \right.$$
$$\left. t \in [0,1],\ \theta \in [0, 2\pi]\right\},$$

where $\sigma = \sigma(u,t) = \frac{1}{2}\ln[(1-t)^4 + 4u^2 t^2 (1-t)^2 + t^4]$ *and* $u > 0$.

Corresponding to the shape of the nondenegerate $\overset{(u)}{\mathcal{C}}_{\{A_0 A_1 A_2\}}$ we classify its
Hopf fibers.

Definition 1.4. The Hopf fibers $\{\overset{(u)}{\mathcal{C}}_{|\mathbb{S}^3}\}$ associated with $\overset{(u)}{\mathcal{C}}_{\{A_0 A_1 A_2\}}$ is
called elliptic, parabolic and hyperbolic according as $u \in (0, 1)$, $u = 1$ and
$u > 1$.

Combining Theorem 1.3 and Theorem 1.4 we get the following

Corollary 1.1. *Let* $C^{exp}_{\{A_0 A_1 A_2\}}$ *be the b-exponential image of a standard Bézier parabola* $C_{\{A_0 A_1 A_2\}}$ *which is described in Theorem 1.3. Then we have the following:*

(1) *Its sphere* $(\mathbb{S}^2\text{-})\text{image } \pi^{-1}(C^{exp}_{\{A_0 A_1 A_2\}})$ *is the curve*

$$
C^{exp}_{|\mathbb{S}^2} : \vec{r}(t) = e^{-\kappa} \left((\lambda - \mu)e^{\mu}, \ \mu(e^{\mu} - e^{\lambda}), \ \begin{vmatrix} \lambda & \mu \\ 1-e^{\lambda} & 1-e^{\mu} \end{vmatrix} \right) \begin{pmatrix} \vec{a}_0 \\ \vec{a}_1 \\ \vec{a}_2 \end{pmatrix} ;
$$

(2) *Its Hopf fibers* $h^{-1}(C^{exp}_{|\mathbb{S}^2})$ *in* $\mathbb{S}^3$ *is*

$$
C^{exp}_{|\mathbb{S}^3} : (x_0, \ x_1, \ x_2, \ x_3)^T
$$

$$
= \left\{ \frac{-e^{\frac{\kappa}{2}}}{\sqrt{2(e^{\kappa}+(1-t)^2)}} \left[e^{-\kappa} \begin{pmatrix} \sin\theta & 0 & 0 \\ \cos\theta & 0 & 0 \\ 0 & -\cos\theta & -\sin\theta \\ 0 & \sin\theta & -\cos\theta \end{pmatrix} \begin{pmatrix} (\lambda - \mu)e^{\mu} \\ \mu(e^{\mu} - e^{\lambda}) \\ \begin{vmatrix} \lambda & \mu \\ 1-e^{\lambda} & 1-e^{\mu} \end{vmatrix} \end{pmatrix} \right. \right.
$$

$$
\left. \left. + \begin{pmatrix} \sin\theta \\ \cos\theta \\ 0 \\ 0 \end{pmatrix} \right] ; \quad t \in [0,1], \ \theta \in [0, 2\pi] \right\},
$$

where

$$
\kappa = \kappa(t) = \frac{1}{2} \ln \left[\begin{vmatrix} (\lambda - \mu)^2 & -\mu^2 \\ (e^{\mu} - e^{\lambda})^2 & e^{2\mu} \end{vmatrix} + \begin{vmatrix} \lambda & \mu \\ 1 - e^{\lambda} & 1 - e^{\mu} \end{vmatrix}^2 \right],
$$

and λ *and* μ *are the functions described in Theorem 1.3.*

Remark 1.4. Statements can be formulated analogously as above for the b-trajectories $C^{(n)}_{\{A_0 A_1 A_2\}}$ of a standard Bézier parabola and their corresponding envelopes $\mathcal{E}_n$ for the sets of polar lines with respect to $\triangle A_0 A_1 A_2$ [2, pp.117-118].

Remark 1.5. All the objects and statements considered in this section have their equivalents in the Lie algebra $(\mathbb{H})$ of real quaternions, if we identify the sphere $\mathbb{S}^3$ with the subspace of the unit quaternions. More exactly, if we take an orthonormal basis $\{\vec{e}_1, \vec{e}_2, \vec{e}_3\}$ of $\mathbb{R}^3$, we have the following identification:

$$
\mathbb{R}^3 = \text{span}\{\vec{e}_1, \vec{e}_2, \vec{e}_3\} \xleftrightarrow{1-1} \{0\} \times \mathbb{R}^3 = \text{span}\{\varepsilon_k = (0, \vec{e}_k), \ k = 1, 2, 3\} = \mathbb{H}^*.
$$

Here, $\mathbb{H}^*$ is the Lie ideal (with respect to the quaternionic commutator $[\alpha, \beta]$) of the purely imaginary quaternions:

$$\mathbb{H}^* \subset \mathbb{H} \equiv \mathbb{R}^4 = \left\{ \alpha = (a_0, \vec{a}) : a_0 \in \mathbb{R}, \ \vec{a} = (a_1, a_2, a_3) \in \mathbb{R}^3 \right\}$$

$$= \mathrm{span}\{\varepsilon_0 = (1, \vec{0}), \varepsilon_1, \varepsilon_2, \varepsilon_3\}.$$

Usually, for algebraic reasons, the geometrically arising quaternionic basis $\{\varepsilon_0, \varepsilon_1, \varepsilon_2, \varepsilon_3\}$ is denoted like $\{1, i, j, k\}$. So, to the Bézier curve $\overset{(u)}{\mathcal{C}}_{\{A_0 A_1 A_2\}}$ the curve $\overset{(u)}{\mathcal{C}}_{|\mathbb{H}^*} : \vec{\rho}(t) = (1-t)^2 i + 2ut(1-t)j + t^2 k \ \in \mathbb{H}^*$ corresponds. The $\mathbb{H}$-analogous curve of the standard cubic Bézier curve $\mathcal{C}_{\{A_0 A_1 A_2 A_3\}}$ is

$$\mathcal{C}_{|\mathbb{H}} : \vec{\rho}(t) = (1-t)^3 + 3t(1-t)^2 \, i + 3t^2(1-t) \, j + t^3 \, k \ \in \mathbb{H}, \quad t \in [0, 1].$$

Remark 1.6. The $\mathbb{S}^3$-image of $\mathcal{C}_{|\mathbb{H}}$ is the curve

$$\mathcal{C}_{|\mathbb{S}^3} : \vec{r}(t) = \frac{1}{|\vec{\rho}(t)|}(x_1, x_2, x_3, x_4),$$

$$x_1 = (1-t)^3, \ x_2 = 3t(1-t)^2, \ x_3 = 3t^2(1-t), \ x_4 = t^3,$$

and the image of $\mathcal{C}_{|\mathbb{S}^3}$ through the Hopf map turns to the following curve over the sphere $\mathbb{S}^2$:

$$\mathcal{C}_{|\mathbb{S}^2} : \vec{R}(t) = \frac{1}{|\vec{\rho}(t)|^2}\left(\vec{\rho} H_1 \vec{\rho}^T , \ \vec{\rho} H_2 \vec{\rho}^T , \ \vec{\rho} H_3 \vec{\rho}^T \right), \quad t \in [0, 1],$$

where H_1, H_2, H_3 are the matrices of the Hopf map in Definition 1.1 and

$$|\vec{\rho}(t)|^2 = \sum_{s=1}^{4} x_s^2 = (1-t)^6 + 9t^2(1-t)^4 + 9t^4(1-t)^2 + t^6,$$

$$\vec{\rho} H_1 \vec{\rho}^T = x_1^2 + x_2^2 - x_3^2 - x_4^2$$
$$= (1 - 2t)[(6 - 2\sqrt{6})t^2 - 2t + 1][(6 + 2\sqrt{6})t^2 - 2t + 1],$$

$$\vec{\rho} H_2 \vec{\rho}^T = 2(x_1 x_4 + x_2 x_3) = 20t^3(1-t)^3,$$

$$\vec{\rho} H_3 \vec{\rho}^T = 2(x_2 x_4 - x_1 x_3) = 6t^2(1-t)^2(2t - 1).$$

2. Quaternionic Hermitian vector spaces and associated Bézier type almost complex structures

2.1. *The right quaternionic Hermitian vector space*

Let $\mathbb{H}^n = \underbrace{\mathbb{H} \times \mathbb{H} \times \ldots \times \mathbb{H}}_{n \ copies}$ be the right quaternionic vector space over the skew-field of right acting real quaternions $\mathbb{H} = \mathrm{span}\{1, i, j, k\}$. A quternionic-linear (q-linear) combination for arbitrary

$$\alpha = (\alpha_1, \alpha_2, \ldots, \alpha_n), \ \beta = (\beta_1, \beta_2, \ldots, \beta_n) \in \mathbb{H}^n$$

with right quaternionic multipliers $\lambda, \ \mu \in \mathbb{H}$ is of the form

$$\alpha\lambda + \beta\mu = (\alpha_1\lambda + \beta_1\mu, \ \alpha_2\lambda + \beta_2\mu, \ \ldots, \alpha_n\lambda + \beta_n\mu).$$

In particular, the right q-multiplication with the conjugates of the basic q-imaginary units $\bar{i} = -i, \bar{j} = -j, \bar{k} = -k$ define the linear operators $I, J, K \in \text{Hom}(\mathbb{H}^n)$. The action of an arbitrary $A \in \{I, J, K\}$ is

$$A : \mathbb{H}^n \longrightarrow \mathbb{H}^n$$
$$\alpha \longmapsto \alpha' = A(\alpha) \overset{def}{=\!=\!=} -\alpha.a$$

for the corresponding small letter $a \in \{i, j, k\}$. These I, J, K satisfy the quaternion relations: $A^2 = -\text{id}$, $AB = -BA = C$ for each cyclic permutation (A, B, C) of (I, J, K). In [3] systematic descriptions of the quaternionic Hermitian vector spaces and structures are given. We shall give some additional observations based on Remark 1.5 and the identification $\mathbb{H} \equiv \mathbb{C}^2$: For an element $\alpha = (a_0, \vec{a}) = a_0 + a_1 i + a_2 j + a_3 k \in \mathbb{H}$ we take $u = a_0 + a_1 i$, $v = a_2 + a_3 i \in \mathbb{C}$. Since we have $\alpha = u + vj$ because $ij = k$, we identify α with $(u, v) \in \mathbb{C}^2$. By this identification, the $\mathbb{H}$-conjugate element $\bar{\alpha}$ of α is expressed as $\bar{\alpha} = \bar{u} - vj$ with the $\mathbb{C}$-conjugate element $\bar{u}$ of $u \in \mathbb{C}$. For given $\alpha = u + vj$, $\beta = p + qj \in \mathbb{H} \equiv \mathbb{C}^2$, we see their q-summation and their q-multiplication which is temporary denoted like "$\bullet$" are of the form

$$\alpha + \beta = (u + p) + (v + q)j \quad \text{and} \quad \alpha \bullet \beta = \begin{vmatrix} u & v \\ \bar{q} & p \end{vmatrix} + \begin{vmatrix} u & v \\ -\bar{p} & q \end{vmatrix} j.$$

If we interpret $\mathbb{H}$ as $\mathbb{R}^4$, these operations are of the following form: For any $\alpha, \beta \in \mathbb{H} \equiv \mathbb{R}^4 = \mathbb{R} \times \mathbb{R}^3$ which are represented as

$$\alpha = (a_0, \vec{a}) = (a_0, a_1, a_2, a_3), \ \beta = (b_0, \vec{b}) = (b_0, b_1, b_2, b_3),$$

we have

$$\alpha + \beta = (a_0 + b_0, \vec{a} + \vec{b}), \qquad\qquad \bar{\alpha} = (a_0, -\vec{a}),$$
$$\alpha \bullet \beta = \left(a_0.b_0 - \vec{a}.\vec{b}, \ \begin{vmatrix} a_0 & b_0 \\ -\vec{a} & \vec{b} \end{vmatrix} + \vec{a} \times \vec{b} \right), \qquad \overline{\alpha \bullet \beta} = \bar{\beta} \bullet \bar{\alpha}.$$

Here $\vec{a} \cdot \vec{b}$ and $\vec{a} \times \vec{b}$ are the geometric scalar and vector cross products.

An element $\alpha = (\alpha_1, \alpha_2, \ldots, \alpha_n) \in \mathbb{H}^n$ can be represented as follows:

$\boxed{\text{i}}$ in $\mathbb{C}^{2n}$, by putting $\alpha_\ell = z_\ell + z_{n+\ell} j$ for $\ell = 1, \ldots, n$, we see it is represented as $\alpha = (z_1, \ldots, z_n, z_{n+1}, \ldots, z_{2n}) \in \mathbb{C}^{2n}$;

$\boxed{\text{ii}}$ in $\mathbb{R}^{4n}$, by putting $\alpha_\ell = (a_\ell, \vec{a}_\ell)$ with $a_\ell = \mathrm{Re}z_\ell$ for $\ell = 1, \ldots, n$, we see it is represented as $\alpha = (a_1, \ldots, a_n, \vec{a}_1, \ldots, \vec{a}_n) \in \mathbb{R}^n \times \mathbb{R}^{3n}$. We put $\mathrm{Re}\alpha = (a_1, \ldots, a_n)$.

Now we consider the q-Hermitian inner product $(\alpha, \beta) := \sum_{k=1}^n \alpha_k \bullet \overline{\beta_k}$ for $\alpha = (\alpha_1, \alpha_2, \ldots, \alpha_n), \beta = (\beta_1, \beta_2, \ldots, \beta_n) \in \mathbb{H}^n$, which is a bilinear form satisfying $(\alpha\lambda, \beta\mu) = \overline{\lambda} \bullet (\alpha, \beta) \bullet \mu = \overline{\mu} \bullet (\overline{\beta}, \overline{\alpha}) \bullet \lambda$, for arbitrary $\lambda, \mu \in \mathbb{H}$. If we put $\alpha_\ell = z_\ell + z_{n+\ell}j$, $\beta_\ell = u_\ell + u_{n+\ell}j$ for $\ell - 1, \ldots, n$, then we have

$$(2) \qquad (\alpha, \beta) = \sum_{\ell=1}^n z_\ell\overline{u_\ell} + \sum_{\ell=1}^n \begin{vmatrix} z_{n+\ell} & u_{n+\ell} \\ z_\ell & u_\ell \end{vmatrix} j$$

in $\mathbb{C}^{2n}$-representation. Here, the first term $\sum_{\ell=1}^n z_\ell\overline{u_\ell}$ is a Hermitian form and the coefficient $\sum_{\ell=1}^n \begin{vmatrix} z_{n+\ell} & u_{n+\ell} \\ z_\ell & u_\ell \end{vmatrix}$ of the second term is a symplectic form. If we put $\alpha_\ell = (a_\ell, \vec{a}_\ell)$, $\beta_\ell = (b_\ell, \vec{b}_\ell)$ for $\ell = 1, \ldots, n$, with $a_\ell = \mathrm{Re}\,z_\ell$ and $b_\ell = \mathrm{Re}\,u_\ell$, we have

$$(\alpha, \beta) = \Big(\sum_{\ell=1}^n (a_\ell b_\ell + \vec{a}_\ell \cdot \vec{b}_\ell)\Big), \ -\sum_{\ell=1}^n \Big(\begin{vmatrix} a_\ell & b_\ell \\ \vec{a}_\ell & \vec{b}_\ell \end{vmatrix} + \vec{a}_\ell \times \vec{b}_\ell\Big)$$

in $\mathbb{R}^{4n}$-representation. Here, if we denote by $\langle *, * \rangle$ the Euclidean scalar product in $\mathbb{R}^{4n}$, the first component, which is the quaternionic real part $\mathrm{Re}^{\mathbb{H}}(\alpha, \beta)$ of (α, β), coincides with $\langle \alpha, \beta \rangle$. The second component can be expressed as $\mathrm{Im}^{\mathbb{H}}(\alpha, \beta)$, which is the quaternionic imaginary part of (α, β). Moreover, we have ([3, p.189, proof of Lemma 2.1])

$$(\alpha, \beta) = \langle \alpha, \beta \rangle + \langle \alpha, I\beta \rangle i + \langle \alpha, J\beta \rangle j + \langle \alpha, K\beta \rangle k.$$

The linear space $\mathbb{H}^n$, endowed with the quaternionic Hermitian inner product, is called quaternionic unitary linear (vector) space and is denoted $(\mathbb{H}^n, (*, *))$.

2.2. Bézier type almost complex structures on a quaterenionic Hermitian vector space

Let $\mathcal{H}$ be the subset which is $\mathbb{R}$-linearly spaned by I, J, K:
$$\mathcal{H} = \mathrm{span}\{I, J, K\} = \{\mathbb{J} = aI + bJ + cK : (a, b, c) \in \mathbb{R}^3 \backslash \{(0,0,0)\}\}.$$
The frame of the quaternionic-like triplet of linear transformations $\{I, J, K\}$ is an orthonormal positively oriented basis for $\mathcal{H}$. It is called canonical (basic) frame. Thus $\mathcal{H}$ and the geometric set of points $\Sigma \backslash \{O\}$ of the euclidean

space $\mathbb{E}_3$ with orthonormal coordinate system $\{O, \vec{e}_1, \vec{e}_2, \vec{e}_3 = \vec{e}_1 \times \vec{e}_2\}$ are (1-1)-equivalent by the linear bijection

$$I \longleftrightarrow \vec{e}_1 = \overrightarrow{OA_1}, \ J \longleftrightarrow \vec{e}_2 = \overrightarrow{OA_2}, \ K \longleftrightarrow \vec{e}_3 = \overrightarrow{OA_3}.$$

We separate $\mathcal{H}$ with respect to the basic frame $\{I, J, K\}$ and consider the following subsets by identifying this frame with the basic $\triangle A_1 A_2 A_3$:

$$\mathcal{H}^b = \{\mathbb{J} \in \mathcal{H} : a + b + c = 1\}, \quad \mathcal{H}^\mathbb{C} = \{\mathbb{J} \in \mathcal{H} : a^2 + b^2 + c^2 = 1\}.$$

We call them the set of the barycentric structures and the unit sphere of the set compatible almost complex structures: $\mathbb{J}^2 = -Id$, respectively, as in Section 1. We set for any $u > 0$ the following 1-parameter subsets

$$\overset{(u)}{\mathcal{H}}(t) = \{\overset{(t)}{\mathbb{J}} \in \mathcal{H}^b : (a, b, c) = \frac{1}{B_2(u,t)}((1-t)^2, \ 2ut(1-t), \ t^2), t \in [0,1]\},$$

$$\overset{(u)\mathbb{C}}{\mathcal{H}}(t) = \{\overset{(t)}{\mathbb{J}} \in \mathcal{H}^\mathbb{C} : (a, b, c) = e^{-\sigma(u,t)}((1-t)^2, \ 2ut(1-t), \ t^2), t \in [0,1]\}.$$

Recalling (1) and Theorem 1.4 (1), we call them the 1-parameter set of Bézier type structures and the 1-parameter set of compatible Bézier type almost complex structures, respectively. Corresponding to the cases $u \in (0,1)$, $u = 1$ and $u > 1$, we call $\overset{(u)}{\mathcal{H}}(t)$ (resp. $\overset{(u)\mathbb{C}}{\mathcal{H}}(t)$) the subsets of the elliptic-like, parabolic-like and hyperbolic-like Bézier type structures (resp. almost complex structures). By means of Remark 1.6 there exists a one-parameter set

$$\mathcal{H}^3(t) = \left\{ \overset{(t)}{\mathbb{J}} = \frac{X_1}{|\vec{\rho}|^2} I + \frac{X_2}{|\vec{\rho}|^2} J + \frac{X_3}{|\vec{\rho}|^2} K , \ t \in [0,1] \right\}.$$

We call its elements standard cubic Bézier type almost complex structures. There also exist barycentric trajectories and b-exponential images of such a structures under the action of the set $\mathcal{A}$ of left stochastic like matrices.

References

1. R.B. BROWN AND A.GRAY, *Vector cross products*, Comment. Math. Helv. **42**(1967), 222–236.
2. M. HRISTOV, *Some geometric properties and objects related to Bézier curves*, Topics in differential geometry, complex analysis and mathematical physics, World Sci. Publ., 2009, pp.109–119.
3. K.TSUKADA, *Parallel submanifolds in a quaternion projective space*, Osaka J. Math., **22**(1985), 187–241.

Received January 31, 2011
Revised   May 11, 2011